TIM SPECTOR

Tim Spector is a professor of genetic epidemiology at King's College London and honorary consultant physician at Guy's and St Thomas' Hospitals. He is a multi-award-winning expert in personalised medicine and the gut microbiome, and the author of four books, including the bestselling *The Diet Myth*. He appears regularly on TV and radio around the world, and has written for the *Guardian*, *BMJ*, and many other publications.

TIM SPECTOR

Spoon-Fed

Why Almost Everything We've Been Told About Food Is Wrong

VINTAGE

7 9 10 8

Vintage is part of the Penguin Random House group of companies
whose addresses can be found at global.penguinrandomhouse.com

Penguin
Random House
UK

Copyright © Tim Spector 2020

Tim Spector has asserted his right to be identified as the
author of this Work in accordance with the Copyright,
Designs and Patents Act 1988

First published in Vintage in 2022

First published in trade paperback by Jonathan Cape in 2020

penguin.co.uk/vintage

A CIP catalogue record for this book is
available from the British Library

ISBN 9781529112733

Printed and bound in Great Britain by Clays Ltd, Elcograf S.p.A.

The authorised representative in the EEA is Penguin Random House
Ireland, Morrison Chambers, 32 Nassau Street, Dublin D02 YH68

Penguin Random House is committed to a sustainable future
for our business, our readers and our planet. This book is made
from Forest Stewardship Council® certified paper.

MIX
Paper from
responsible sources
FSC
www.fsc.org   FSC® C018179

To Juno

CONTENTS

PREFACE

In March 2020, only a few days after the first edition of *Spoon-Fed* went off to the printers, all our lives changed. When the first wave of COVID-19 hit London, my research department was shut down and we were all sent home by the university. Cycling home from work that day, feeling depressed, I came up with the idea of converting the nutrition app we had been developing with the data science company ZOE into a free app to fight COVID. Luckily my colleagues at ZOE enthusiastically agreed, and within five days the team, with help from academic colleagues, had a working version ready to be rolled out. The app was an instant hit on social media and was downloaded a million times within 48 hours. Despite various attempts by the government to stop it in the early months, 18 months later it now has nearly five million users in the UK, US and Sweden, making it the world's largest citizen-science project.

The success of the app was due to several factors: firstly, people wanted desperately to tell personal stories of their COVID-19 symptoms when no one else was listening and they couldn't even speak to a doctor; secondly, they wanted to be part of a larger community effort helping others; thirdly, unlike other surveys, they wanted to have regular feedback in the form of independent information they could trust, not just what the government wanted them to believe. The ZOE COVID-19 study app has stuck to these principles, and has continued to provide up-to-date independent information as the virus evolves. This book is born of the same principles: taking a hard look at the data on food and nutrition and being prepared to provide advice contrary to what governments, food companies, doctors and out-of-date science might tell us.

Thanks to the ZOE COVID-19 study app, we now have data from the world's largest ever diet survey. We now know that the type and quality of food we eat has an impact not just on obesity, but the likelihood of being infected, hospitalised or dying from the virus. There is also clear evidence of an association between gut health and COVID outcome: those who eat more plants have less severe illness. So the COVID-19 pandemic, with its clear associations with obesity, social deprivation and poor gut health, combined with its impact on food security, may have created the perfect storm for change. We now know much more about the connection between diet and the immune system, making an even stronger argument that a good quality diet should be a basic right for everyone – especially for children.

In 2021, an independent report called the UK National Food Strategy (on which I was an advisor) was published. The report shares many of the wider recommendations of this book, from the impact our food choices have on the planet to the urgent need to address our food environment by tackling obesity and educating children about real food. I hope the government has the guts to take the decisive action needed, although, at the time of writing, the signs are not good. The UK government's half-hearted response to the obesity crisis

shows there's still an enormous amount of work to change official mindsets. We may have to continue to rely on a bottom-up
approach through education and spreading the word. When it
comes to food, people do not freely 'choose' their behaviours.
What we eat is also shaped by how our social, political and
economic system is organised – whether healthy food is available, whether it is affordable, and whether we have the luxury
to care. In other words, healthy eating is not something we can
do solely on our own. It is something that we need to do as a
society. There are signs of growing public awareness of the key
issues, as we saw in the reaction to the government withdraw-
ing free school meals for deprived children in 2020.

The relationship between food and the environment is
now a mainstream topic, and as we gather more data my views
are also changing. I think we need to take a closer look at the
wider impacts of all animal products on our health and on the
non-human environment. Dairy alternatives have come a long
way with big improvements in many plant-based milks that are
made with simple ingredients, making oat milk my personal
milk of choice. Even nut-based probiotic vegan cheeses, which
I once sneered at, are improving in quality and flavour. These
alternatives make cheap, subsidised dairy consumption from
huge, badly treated dairy cattle herds increasingly difficult to
defend in terms of morality, flavour or health. The benefits of
eating fish have also been greatly exaggerated, and if we care
about the survival of our planet we need to give the sea a break
to recover and reset. I still believe that veganism is not necessarily the best diet, but it is irrefutable that the more your
diet is made up of plants, the healthier you and your planet
will be. Consuming ethically produced animal products as an

occasional treat to supplement a diet rich in nuts, vegetables, pulses, unrefined grains and fruits, much as our ancestors did, seems to be the way forward, perhaps more so than growing test-tube meat.

Since this book was first published, I have been pleased to see a growing – and long overdue – focus in science and in the media on the dangers of ultra-processed foods (UPFs). I hope everyone who reads this book will be able to understand the difference between a modified food such as canned beans, which can be beneficial, and an ultra-processed food such as a supermarket seeded loaf, which can be loaded with chemical ingredients despite appearing to be 'healthy'. In the UK and US we currently consume over half our daily food calories in UPF form, compared with under 10 per cent in Mediterranean countries. After thirty years of silence and a virtual unofficial ban on UPF studies, some trials are defying the food industry and finally showing how, regardless of calories, UPFs alter our metabolism and induce overeating, especially in children. We need to return to whole foods to prevent damage to our bodies and brains, especially for children where the blueprint for our lifelong health is set. The overeating of these artificially reconstituted foods is the biggest health threat we face, and it needs urgent action.

We all reacted differently to COVID-19. I totally lost my appetite for the three weeks when I was ill, but I was lucky. Millions lost their sense of smell and taste for months, with devastating effects on their enjoyment of food. Our diet survey of a million people in 2021 showed that improving food choices was a happy outcome of COVID-19 lockdowns for about a third of the population, who found they cooked more meals from scratch, baked bread, ate food as a family and reached a healthy weight as a consequence. In contrast, a third of the population found themselves eating more unhealthy takeaways and snacks and drinking excessive alcohol, leading to weight gain. These results were not predicted by income or social class, showing how we can all change our habits when our environment changes.

The pandemic caused mood changes in most of us (myself included), and research into the connection between food and mental health has been developing rapidly. We now understand better how food, mood and microbes are connected and how we can start to manipulate our mood through diet. The implications are astonishing: from reducing postprandial inflammation caused by sugar and fat spikes in the blood, to eating foods that are helpful in staving off dementia, to avoiding UPFs that may be accelerating mental health problems. The evidence all points to a similar beneficial dietary pattern of plenty of colourful plants, nuts and legumes and probiotic rich fermented foods to halt the mental health pandemic.

Following on from our pioneering PREDICT nutrition studies, many people have now used ZOE home-testing technology and have been able to discover exactly how food has an impact on them on a personal, meal-by-meal level. Since this book was first published, this has now generated the biggest food-microbiome database in the world, and as the data grows, so will the fine detail of the personalised advice.

While we all need to change our food system to give healthy food the priority it deserves, increasing numbers of us can now use the latest technology to understand our body's responses to every meal. All of us now have the power to change our health and happiness with the latest scientific knowledge. There is no longer the need to rely on a creaking healthcare system, government messages or poorly informed medical staff. The future of high quality and personalised nutrition is here and there is no room for misinformation on our plates.

Tim Spector
November 2021

INTRODUCTION

Most of us learn our first food myths as children. When I was young, I was told that some special foods would make me grow quickly (milk and cereal), make me brainy (fish), give me acne (chocolate), or give me big muscles (meat and eggs). I was encouraged to eat spinach because of Popeye, but never told anything about the benefits of lentils, broccoli or beans, and I was told nuts were unhealthy snacks because of the cholesterol. I was also told I would be ill without a proper breakfast. My mother, brought up in the war years, told me virtually no food was too mouldy to eat, and leaving food on your plate was unacceptable. I don't recall having any 'proper' meals that didn't involve meat or fish. Vitamins were seen as very important, especially vitamin C, which was taken as supplements or drunk as orange juice. Other unquestioned advice included never swimming within an hour of eating, never eating just before bedtime and the importance of exercising to lose weight. Not one of these ideas is backed up by science, and many of them turn out to be categorically wrong, but they were repeated so often that I still find them hard to unlearn as an adult. We all inherit similar notions, and opinions about food – whether well-intentioned or not – only multiply as we get older.

Eat less fat. Eat less sugar. Eat your five-a-day. Eat more starchy vegetables, never skip meals, eat little and often, drink at least eight glasses of water a day, drink less caffeine, drink less alcohol, eat less meat and dairy, eat more fish, use vegetable oils not butter, count your calories and switch to diet drinks. We have become used to being told how, when and what we should be eating. These messages come from many different sources: national guidelines, the media, advertising and even food labels and cereal packets, as well as posters and leaflets in hospitals and doctors' surgeries. With all this advice, we should surely all be healthier, slimmer and free of diet-related illnesses. Instead, since 1980, rates of obesity, food allergies and diabetes in most countries have rocketed, along with unexplained rises in dementia. Despite advances in treatment, rates of heart disease and cancer are on the rise, and the recent increase in longevity has flattened off and is showing signs of decline. Faced with a myriad of food choices and a tidal wave of misinformation, many of us want a simple quick-fix solution. Even the most cynical can find themselves absorbing unfounded advice with simplistic messages. We too readily fall for the claims of lifestyles such as clean-eating, vegan, ketogenic, high-fat–low-carb, Palaeolithic, gluten-free or lectin-free, or the myth of vitamin supplementation. The belief and confidence of the advocates of these diets and their followers can be very convincing.

My scientific research has focused increasingly on nutrition and food-related questions in recent years. I have been astonished to discover how much of what we are told about food is at best misleading, and at worst, downright wrong and dangerous to our health. As we will see, this is true whether the advice comes from dieticians, doctors, government guidelines, science reporting or anecdotally through friends and family. How did we get in this mess where unqualified people dictate the best ways for us to eat? This situation is unique in the field of medicine and science. There are many reasons for this, but I would pinpoint three major obstacles to better understanding about food and nutrition: bad science, misunderstanding of the results, and the food industry. Diet is the most important medicine we all possess. We urgently need to learn how best to use it.

Science is complicated. The study of food and health nutrition is one of the newest sciences, and only appeared in most countries from the 1970s in response to growth in the processed food industry and the desire for governments to advise on avoiding nutritional deficiencies. In most countries, nutrition is still not seen as the domain of medicine and the two areas of sciences rarely overlap, with few medics ever studying nutrition or vice versa, so the experience, methods, trials and errors made in testing pharmaceutical medicines and dealing with the food industry have not been fully shared with the nutrition scientists. Despite the fact that it wrestles with some of the most important questions of our time, academic nutrition is seen as one of the least glamorous or important areas of science. I work closely with a commercial nutrition company ZOE, which has recruited excellent analysts who began their careers in the supposedly more glamorous fields of astrophysics, maths and economics before transferring to work on large-scale food data. But most nutritional experts, with a few exceptions, remain isolated and feel unloved and undervalued by their universities and funding bodies, which are largely sponsored by the food industry. Instead of performing the clinical, large-scale studies we desperately need, they are forced to spend most of their time teaching or performing small-scale short-term studies on foods.

Let us be clear: doing good research on food is difficult, and funding has been woefully inadequate for the massive long-term studies that are necessary to test one food or diet against another in humans. It costs nearly $1 billion to bring a new drug to market; yet we spend only a tiny fraction of that amount when we assess foods or diets. For this reason, most of what we have been told about the benefits or risks of foods come from either dubious test-tube studies or small-scale studies of rodents given artificial diseases that are rarely relevant to humans. Almost every day there is a new example in the media. A typical series of headlines, in 2019, announced that eating walnuts daily protects against cancer and colitis. In fact, the scientific paper merely described that mice who had been given chemicals to mimic human disease showed a slight improvement in metabolic profiles after two weeks of walnut therapy.[1] The study was small and appeared in a modest but genuine nutrition journal, but the sponsors of the study – the California Walnut Commission – must have been delighted with the free publicity. These studies are next to useless, particularly as many other similar mice experiments are performed relatively cheaply but never report findings when they don't provide the 'right' result for the funders.

Scientific research has improved, and we have come to rely more on large-scale observational studies following tens or hundreds of thousands of people over many years. These have provided important insights but tend to be based on simple but often unreliable questionnaires. The instruments used to collect diet data have been crude, with overweight people prone to routinely under-reporting consumption and skinny people over-reporting. In general, most people under-report eating foods perceived as unhealthy. New technology involving smartphone cameras and apps is rapidly reversing this. A highly critical summary of the nutrition field and these observational studies in 2018 noted many flaws, including the point that positive findings are routinely over-reported. In a giant meta-analysis combining all of these kind of studies (e.g., eggs, dairy, refined grains, legumes etc.), all twelve food groups studied were associated with either increased or decreased risk of death.[2] Of course, this is highly improbable, but such results further encourages an unrealistic good/bad food dichotomy to which we are all susceptible.

When you look at hundreds or thousands of possible associations between foods and diseases, you are bound to find spurious links. Nutrition studies are much harder to perform reliably than drug studies, and a research framework to assess them differently to drugs studies was only proposed for the first time in 2019.[3] Applying the strict criteria used in drug studies to food has led to some spurious conclusions. In 2019, a Canadian research group hit the headlines with the news that it was OK to eat meat after all. It turned out that in their summary of the data they had excluded half the available studies and received large undisclosed grants from the food industry. They also wrote something similarly controversial about the lack of harm of sugar two years earlier.[4] Science has oversimplified food in the same way that we used to view genetics twenty years ago. Early gene studies I was involved in found hundreds of possible links between large stretches of genes and diseases using hundreds of markers. We 'discovered' many new genes for, say, obesity, ageing, osteoporosis or diabetes. These studies produced a good deal of publicity, which was great for my career as a scientist, but they mostly turned out to be rubbish. New gene chip technology has revealed the full complexity of our genes, and showed that what we call a 'gene region' often contains 200–1,000 completely different genes we couldn't previously detect. So the idea that a single gene could be discovered for any common disease or health condition was shown to be a myth. Some of these so-called discoveries were sold for hundreds of millions of dollars but were next to useless. Today's equivalent myths about food that appear to be science-based often come from primitive test-tube studies. In these, human or mouse cells are developed and exposed at very high doses to a single chemical contained in some food or released when a product is heated or cooked. Nearly every substance tested this way has been shown to be 'unsafe', i.e., at least slightly carcinogenic. The food industry uses the reverse technique to show in small studies that its products are safe or beneficial. Most foods contain thousands of chemicals, and we are never exposed to a single one in this artificial isolated way; so even if the results are reliable and repeatable by other groups (which they often are not), the conclusions are always dubious.

Part of the problem is the way food science is based on a centuries-old misunderstanding that divided our food into just three major subgroups: carbohydrates, fats and protein. These three groups were seen as sources of energy as calories, which needed to be eaten in the right proportions to prevent deficiencies (and, as we shall see, calories are themselves a flawed and hopelessly unreliable measure of anything). But this division of all foods into one of three groups is like classifying all humans as African, European or Asian, and then recommending standard treatments and finding differences in health, strength or intellect according to these crude categories. The idea that we can separate, say, carbohydrates and protein, as so many diet advocates, doctors and government guidelines recommend, is scientific nonsense. All foods are a complex mixture of carbohydrates, fats and proteins. When the science itself is dangerously oversimplified and misleading, dumbing it down even further into rules and guidelines only makes it more likely that messages will get distorted.

The problem doesn't just lie with the science; just as big a problem is the way the results are misinterpreted and misunderstood. Studies often produce hundreds of results, and the interesting findings and their risks will always be picked up by eager journalists and turned into shocking but often misleading headlines. The idea that you can take one longitudinal population study that shows eating two rashers of bacon per day increases the risk of heart disease and death is one thing. But to extrapolate from this that it will reduce your lifespan by a decade is ridiculous – this is greater than the health risk of regular smoking. Similarly, some health foods are promoted in outrageous ways – we are told that eating a daily handful of one type of nut or berry can increase our lifespan by fifteen years. Having two small glasses of wine daily may increase your relative risk of one type of cancer by, say, 10 per cent (compared to a non-drinker), but the personal risk of developing that cancer is probably less than one in ten thousand. Few people can be expected to disentangle the different ways that such risks are presented to us.

But the problem goes far beyond spurious headlines, and this simplified or misleading science often forms the basis of government guidelines. Governments began to tell citizens what to eat during rationing in the Second World War, when food resources were stretched and the government needed armies of healthy citizens. Obesity was extremely rare, and the greatest challenge to public health was malnutrition, so the government handed out advice to help avoid vitamin deficiency. The early success of this approach set the tone for the next sixty years, and established the assumption that health problems could be resolved by altering one key component of diet, such as adding vitamin C or reducing fat levels, because population studies had shown these components were linked with disease. Fat became the bogeyman for decades and people were encouraged to eat more carbohydrates and protein instead, leading to the creation of low-fat, highly processed foods. Even now that the fat hypothesis has been conclusively challenged, a new single villain in the form of sugar has been put up to replace it, spawning multiple low-sugar processed foods. As we demonised one food, we never asked, 'what do we replace it with?' As we fiddled with percentages, we forgot about healthy groups of food. We were told to eat more often and so ate more tempting snacks and even more highly processed, low-fat foods, and we fed them to our children. As a result, we got fatter and sicker.

Judging a food by any of its single components is another problem. Fructose is a common sugar found in many fruits and just one of over six hundred chemicals in a banana, which some people say we should avoid because of its high fructose content. Another chemical currently vilified by some is lectin, a protein found in uncooked beans that is toxic to humans. But they ignore the fact that the plants with the most lectin, such as beans, lentils and nuts, contain thousands of other healthy chemicals key to the best diets on the planet. Plants turn out to be far more complex than we had imagined and many of the chemicals are protective substances called polyphenols (which used to be called antioxidants), which we now know play a key role in our health in fighting cancer and other diseases. The importance of polyphenols has long been overlooked, as they don't work directly on our bodies. In fact, we are unable to make use of them at all without help. That help comes from an organ only recently discovered. The gut microbiome.

Research into the microbiome has shown how reductionist our view of food has been for so many decades. It is not an organ in the conventional sense, but a community of tiny organisms which together weigh as much as our brain. The microbiome consists of a mix of up to 100 trillion bacteria, fungi, parasites and 500 trillion mini viruses, outnumbering the number of cells in our body. The vast majority live in our large intestine, alongside most of our immune cells. Each microbe is capable of producing hundreds of chemicals, which act as mini-factories regulating our immune system, providing many of the key metabolites and vitamins in our bloodstream, including brain chemicals that can affect our mood and even our appetite. Unlike other parts of our bodies, the mix of gut microbes, their genes and the chemicals they produce is unique to each of us and differs in each individual, even genetically identical twins.

This new extra organ has made us realise that thousands of food chemicals interact with thousands of different microbe species to produce over 50,000 chemicals that affect most aspects of our body. When we consume food, it is as much for the benefit of our gut microbes as it is for us. So the way a food affects our body might differ substantially from one individual to another. As yet, there are far too few expert advisors in the field of the microbiome, with no medical specialists, nutritionists or dieticians trained in it. With its mixture of genetics, microbiology, computing and biochemistry, the field of the microbiome is perceived as a daunting subject and a risky, lonely and unsupported career move for nutritionists. Sadly, the people who supply us with food advice have also been far too slow to get up to speed with the new science, hoping that it will prove to be another passing fad.

The assumption we are all identical machines, and that we all respond to foods in the same way, is the most prevalent and dangerous myth about food. It is the basis of all so-called diet advice. And it's not just about our different populations of microbes. As I discuss in chapter 1, normal people can vary tenfold in their blood sugar responses to identical foods. We all respond differently to the same foods and so the idea we can all follow the same advice and calorie limits no longer makes sense, in the same way that we wouldn't all be comfortable in a standard car seat without adjusting it, just because it was designed for an average person. And, while we're on the subject, determining our food needs, such as how many calories to consume a day, by gender is also daft. It has been a deliberate policy of the food industry to ignore or downplay our individual metabolisms, food responses and unique microbes, partly because marketing works better with a simpler message, but also because they especially want to avoid scrutiny or extra tests on the safety of added food ingredients against our gut microbes.

This brings us to the greatest obstacle of all when it comes to dangerously inaccurate food information: the food industry. My scientific research has opened my eyes to the astonishing and pernicious influence of the industry. Until recently, I had no idea of the massive scale, unlimited finances and power of a handful of companies over all of us, and one of my hopes in writing this book is to make more people aware of the situation. While we need to give these companies credit for their ability to feed the growing population and produce increasing varieties of low-cost food that people like to eat with lower rates of spoilage and long shelf lives, they have rapidly become too powerful. Companies such as Nestlé, Coca-Cola, PepsiCo, Kraft, Mars, Unilever each have revenues larger than half the countries in the world; the ten biggest food companies control 80 per cent of store-bought products globally, each generating an average of over $40 billion in annual revenue in 2017[5] and combined profits of over $100 billion in 2018. These global conglomerates took off in the 1970s thanks to the rise of supermarkets and long-shelf-life processed foods, as well as their ability through advertising to send marketing messages into our homes, particularly via the TV. In the 1980s, vitamin fortification of processed foods continued to rise, and products that boasted reduced fat, sugar and salt flew off the shelves. The food industry was delighted to influence and then follow the advice of nutrition expert committees and produce low-fat, low-cholesterol, low-sugar, low-sodium, high-protein heavily processed junk versions of foods. These could be made more cheaply than the original natural products, with ever higher profit margins, longer shelf lives and expanding global markets.

The added bonus was they could now market nearly any ultra-processed junk food as an approved healthy alternative by adding a bright label showing 'low-fat' or with 'added vitamins', accompanied by a whole slew of health claims. Just look at how clever marketing conned us into believing that artificially coloured breakfast cereal, which mainly consisted of sugar or which contained chunks of marshmallows or chocolate, could be (and still is) seen as a healthy meal for children rather than a confectionary. Yoghurt is one of the most microbially rich, healthy foods you can eat. Yet in most countries now it is hard to find a yoghurt that isn't ultra-processed or contains a low-fat synthetic alternative with extra sugar, fake fruit or artificial flavours. All will have some health claim on the label. Snack bars packed with sugar are now labelled as healthy just because they contain small amounts of fibre or protein or some vitamin you don't need. Microwaveable ready meals with over twenty ingredients now carry misleading healthy low-calorie or low-salt stickers, while diabetes-inducing smoothies and juices claim to be supporting 'five-a-day' guidelines.

Clearly, the companies that dominate the food industry are doing so well that they want to keep things just where they are, and are happy to pay for it. As the giant food and drinks companies are merging and increasing in size and power, many people are putting their trust in smaller local operations with clearer perceived ethical values, and shopping less at giant retailers. But as multinationals buy up smaller organic and ethical food companies at a frightening pace (such as Whole Foods by Amazon), it is increasingly hard to know who the good and bad guys are, and who to trust. They love the current guidelines based on general dietary proportions, as they give them great flexibility and distract from the steady rise of ultra-processed foods. The food and drinks industry spends hundreds of millions of dollars on political lobbying to ensure that its national market and interests are protected. In 2009, the top companies declared they paid lobbyists over $57 million in fees in the US alone.[6] This money is spent influencing health officials, and they often get to sit on the expert guideline committees, and nearly always influence the politicians who translate these reports for the public. They affect the committees in other subtle ways; most of the scientists drawing up the guidelines are paid via personal consultancies or receive research grants from food companies, which doesn't mean they are necessarily biased, but does perhaps allow them to be more easily manipulated.

Importantly, the food companies also set the agendas for research. In the US, the food industry provides 70 per cent of the funding for food research, and the picture is similar in other countries. Companies promoting sugar or low-fat foods provide generous grants to academics to focus on areas that suit the industry, such as the benefits of low-calorie foods, exploring exactly how saturated fat in your diet is bad for you, or how lack of exercise (rather than poor diet) is the main cause of the obesity epidemic. This clever plan distracted the field for decades from addressing the real problem of additive-rich, ultra-processed food, meaning that poor-quality, harmful foods such as processed meats have continued to be eaten in vast quantities. In just the same way, the tobacco industry was able to distract us from the real science in the 1960s and 1970s. These successful tactics meant the first proper clinical trial of the harms of junk food versus unprocessed food was only performed in 2019.[7]

The other trick the food industry learned from pharmaceutical companies is that they can influence the key nutrition practitioners with gifts, conferences and selected information, as well as by funding their professional organisations. Like the big pharma companies, the industry encouraged misinformation with small inconclusive studies into aspects of the safety of a product such as artificial sweeteners. Food companies also pay advocates and influencers to cast doubt on larger, more definitive studies that they don't agree with, and use corporate lawyers as well as their massive advertising budgets to punish any rebels. It is hard to be an active nutrition researcher performing expensive clinical studies without encountering people who want to help or influence you. I am no purist: I accepted money from pharma companies to perform trials over ten years ago, and I have taken money from Danone for our research into yoghurt and gut health, as, without this financing, the research would not have happened. So I realise I too am not exempt from possible bias. It may have been a coincidence but three weeks after publishing a critical editorial in the *BMJ* on breakfast advice,[8] I was approached informally by Kellogg's to see if I wanted to act as an advisor on their gut research programme (I said no). Academics like me can feel like Davids pitched again the food industry Goliaths, with their billion-dollar research funds.

In the 2000s, a few people started questioning the orthodoxy that saturated fat in our diet is our main problem. At the time, these critics were widely labelled as fanatics with their own agendas who were selling diet plans, articles or books (which they sometimes were). But in other fields, scientists and officials do admit mistakes. An example is when we were told around 2000 that the data showed that diesel-fuelled cars were better for the environment. In 2018, governments reversed the advice and announced we should switch to petrol or electric cars. They openly admitted mistakes had been made, and it transpired the German car industry and its lobbyists had supplied much of the false information. With nutrition, it was a different story. The establishment was prepared neither to admit that mistakes had been made nor that changes were needed. On top of that, they thought it quite normal to involve the food industry and other interested parties first of all in the discussions of the science, and then secondly in the translation of those findings into firm messages for the public. This was a process that could take years. The longer the change took, the greater the confusion: more questions arose around the science and specific foods were singled out more frequently as potentially harmful. Meanwhile, ultra-processed food was targeted far less often – and so the more the industry was winning.

But things are starting to change. Although this book is structured around some of the most deep-rooted and dangerous myths about food, there is reason for hope. At a nutrition conference in Zurich in June 2018, I witnessed a tipping point. The academic meeting, organised by the *British Medical Journal* and a multinational life insurance company, brought together nutrition experts from around the world, and during the day I sensed many different segments of the healthcare profession were openly contesting nutritional dogmas. General practitioners had patients with type 2 diabetes who were controlling their disease without drugs by following a low-carbohydrate, high-fat diet with initial calorie restriction. This was supported by randomised trials and was in complete contradiction to official incentives that promote drugs first and guidelines recommending that diabetic patients should especially avoid fat. Clinicians were accepting that many of the cornerstones of our philosophy about eating well are based on flawed studies conducted many decades ago. Studies were for example now showing that 'proven' treatments, such as salt restriction in patients with diabetes, was actually increasing risk of death. Respected epidemiologists were also reporting large observational population studies from developing countries that diets high in saturated fats actually protected people from heart disease and diabetes, not the reverse. There was continuing evidence from large, long-term trials that low-fat diets were performing worse than higher-fat Mediterranean-style diets, showing that what else was on your plate was more important than how much fat you ate.

At this Zurich meeting, I presented early data on the enormous individual differences between people in how our bodies react to food, making detailed national guidelines designed to apply to everybody illogical and flawed. Nutrition experts from world-leading institutes such as Harvard and Tufts in the US, who had largely created the original guidelines, were now admitting changes were needed. Bodies in other countries, including the UK, may be more stubborn. Nevertheless, even resistant officials, committees and food industry lobbyists won't be able to stop the tide, as increasing numbers of respected experts are now calling for change.[9] For the first time, scientists like myself can openly challenge some of the diet myths handed down over the past few decades without being ridiculed, vilified or ignored. We have been so distracted by the debate about whether these beliefs about macronutrients or individual foods are true or not, or whether there is yet another truth. Now if we want to, we can open our eyes and see the bigger picture.

I am a scientist and a medical doctor. Yet, over the last decade, I have been shocked by what I have discovered and am still uncovering. I have now revised my opinions on most aspects of food and health that I had learned the conventional way. My last book, *The Diet Myth*, focused on the myths surrounding specific diets and introduced the microbiome. My research has now forced me to look much more widely and deeply at the whole subject of food. This book is born out of an urgent need to rethink the way we eat, to ask better questions and demand higher standards of science and reporting. As we will see, nutritional research is one of the fastest-moving fields of science today, and this book also draws on the very latest science, including the pioneering work of my fantastic team at King's College London and collaborators around the world. As food choice is incontrovertibly linked to our environment, this is no longer just important for our own sake, but for the sake of our planet and future generations. Food science lags behind other disciplines, but at this key point in our history could end up being the most important. Within the past decade, I have changed my mind on most of the subjects in this book, including diet drinks, veganism, eating fish, caffeine, vitamin supplements, pregnancy advice, organic foods and the effects on the environment, and maybe you will too. We all face endless complex food choices every day, and are looking at the prospect of an overcrowded and overheating planet where half the population are obese. There are no simple black-and-white answers. Realising where and how we were conned should help us to get back on track. That's why we all need to rapidly learn more ourselves about the food we eat every day and the science behind it,[10] so we can avoid the smokescreens, and make more informed individual choices.

1.
IT'S PERSONAL

*Myth: Nutritional guidelines and diet plans
apply to everyone*

We humans are complicated: multiple factors influence our health. There are things we can't change, like age or genetic make-up, and things we can, such as our choice of food and drink. And then there are the trillions of bacteria that live in our guts – collectively known as the microbiome – which have a significant impact on health and digestion. The foods we eat are mixtures of many nutrients that affect the body and microbiome in different ways, so unravelling the relationship between diet, metabolism and health is no simple matter.

We are used to listening to government advice and guidelines about nutrition and health and wellness. These guidelines influence not only the general public but the healthcare provided to us by doctors and other health professionals. But can the same health advice really apply to a population of millions of individual humans, with their own lifestyles and unique physiology? Is the one-size-fits-all approach an appropriate base for healthcare policy? We evolved to be omnivores and across the planet eat a huge range of different foods to stay healthy, from Eskimos to African hunter-gatherers to over a billion Asian vegetarians. In our increasingly culturally and ethnically mixed world, is it really possible to say one particular diet suits all people?

In the 2015–20 USDA guidelines, which are the basis for the advice of many other countries, there's a graphic of a plate showing the ideal proportions of a healthy diet. These are 39 per cent from fruit and vegetables, 37 per cent cereals (bread, rice, pasta, potatoes, etc.), 12 per cent proteins from beans, pulses, eggs, meat and fish, 8 per cent dairy and milk and 4 per cent from fatty and sugary foods. We are also told to have five portions of fruit and veg per day, including one glass of fruit juice or smoothie, and fish twice a week, and have 2,000 calories per day for a female and 2,500 for a male.[1] In the UK, the advice is similar, with additional tips, such as never skipping breakfast and drinking eight glasses of water or other fluids per day.[2] The guidelines advise eating little and often, and avoiding large meals in the evening. The US has stricter guidelines than most on reducing saturated fat below 10 per cent of daily intake and reducing salt intake below 2.3 grams of sodium a day (about a teaspoon). For those that choose the alternative routes of the diet and wellness gurus and follow the path of gluten-free, ketogenic, low-carbohydrate, Paleo or intermittent fasting, the problems are the same. Can all these recommendations be right for everyone?

New studies add another layer of complexity, showing that foods with comparable nutritional profiles can have very different effects on health outcomes and the gut microbiome. Some US collaborators of ours asked thirty-four healthy volunteers to collect detailed records about everything they ate over seventeen days, mapping this information against the diversity of microbes in daily stool samples.[3] As expected, although there were several foods that were eaten by most of the participants such as coffee, Cheddar cheese, chicken and carrots, there were plenty of food choices that were unique. While each participant's food choices had an impact on their own microbiome, with certain foods boosting or reducing the abundance of particular bacterial strains, there wasn't a straightforward correlation between them that carried over to other people. For example, beans boosted the proportion of certain bacteria in one person but had far less effect in another.

Although very closely related foods (such as cabbage and kale) tended to have the same impact on the microbiome, unrelated foods with very similar nutritional compositions had strikingly different effects. This tells us that conventional nutritional labelling is not the best way of judging how 'healthy' a food is likely to be. The microbiome is probably the hottest topic in nutrition and health right now, with scientists keen to map and manipulate our bacterial friends, but it's not the whole story.

My team at King's College London are collaborating with researchers at Massachusetts General Hospital, Stanford University in California and the precision nutrition company ZOE.[4] We're running PREDICT – the largest nutritional-science study of its kind anywhere in the world – with the aim of unpicking the complex interacting factors that affect our unique responses to food, especially the regular peaks in sugar, insulin and fat levels in blood that cause metabolic stress and are linked long term to weight gain and disease, and to appetite. We've been studying personal nutritional responses to food initially in 2,000 volunteers from the UK and US, including hundreds of pairs of twins, measuring their blood sugar (glucose), insulin, fat levels (triglycerides) and other markers in response to a combination of standardised and freely chosen meals over two weeks. We also captured information about activity, sleep, hunger, meal timing and frequency, mood, genetics and (of course!) the microbiome, adding up to millions of data points including over 2 million blood glucose measurements from the stick-on monitors (called CGMs) over 130,000 meals and 32,000 specially made muffins. The initial results published in *Nature Medicine* came as a big surprise.[5]

We discovered that individuals have repeatable, predictable nutritional responses to different foods, depending on the proportions of protein, fat and carbohydrates. But importantly there were wide variations between people (up to tenfold), making a mockery of 'averages'. This even included differences between identical twins, who are clones sharing all their genes and much of their environment. Less than 30 per cent of the variation between people's sugar responses is due to genetic make-up, and less than 5 per cent for fat. Unexpectedly, based on previous beliefs, there was only a weak correlation between the two – having a bad response to eating fat couldn't predict whether someone would be a good or bad responder to sugar. Of the thousands of people we tested with identical meals so far, a proportion were close to average for one response, but less than 1 per cent was *exactly* average for their response to all three – sugar, insulin and fat. This means that 99 per cent of us don't conform to some artificial average. We also discovered that identical twins shared just 37 per cent of their gut microbe species. This is only slightly higher than the percentages shared between two unrelated people, underscoring the modest effect of genes. We found that the crude composition of the food you find on the label only explains about a quarter of the metabolic response, and most of the differences were due to unique individual factors, which include the microbiome and genes, but also include the effects of different circadian rhythms in our body clocks, exercise, sleep, and other factors we are still working to uncover.

The wealth of data from PREDICT is being exploited by a large team of academics worldwide, and using machine-learning algorithms, the company ZOE that I am helping has launched a smartphone app to predict how any individual will respond to any food, based on the algorithms and their own personal information. This will help people make healthier choices. The scientific studies are continuing by recruiting thousands more US and UK volunteers for expanded home-based studies. The more people who join the studies, the more data is generated to improve the predictions, which even at an early stage are already 75 per cent accurate, which is much greater than standard clinical tests.

Like many doctors of my generation, I followed official advice on healthy living in middle-age: I didn't smoke cigarettes, tried to exercise regularly and cut back on fats. My breakfast was a low-fat, high-carbohydrate feast of muesli, semi-skimmed milk, wholegrain toast, and a glass of orange juice with a tea or coffee. Recently, I tested my glucose responses to my old 'healthy' breakfast using the new continuous blood glucose monitors (CGMs) as part of the PREDICT study. My blood sugar (glucose level) rose sharply from a resting level of 5.5 to 9.1 mmol and produced a surge in insulin to get my levels back to normal an hour later. I asked my wife to be a guinea pig and have an identical breakfast. Her blood sugars started lower than mine at 4 mmol, but hardly moved above 5.7 mmol.

Our bodies are programmed to absorb the glucose from carbohydrates in food as useful energy and either use it immediately or store it in our muscles or fat cells for later use. Having a high sugar level in our bloodstream is not good for us for more than a few minutes, and our body tries to get rid of it rapidly, mainly by secreting a hormone called insulin. Having regular peaks of blood sugar, insulin or the blood fat triglyceride stresses your system in the long term and promotes the storage of the energy in your fat cells.[6] My body was clearly working harder to make insulin and clear the sugar. I then went and tested (several times) my response to my go-to lunch for ten years in the hospital, a healthy-looking brown-bread tuna and sweetcorn sandwich. The results were worse than I feared, with regular peaks of 10–11 mmol, while again others like my wife had much smaller increases. But I did better than her when eating pasta or basmati rice, which suggests I might not have gained the 10 kilos I did if I had eaten an Italian or Indian lunch instead. I also discovered that compared to other people, grapes, which I also used to eat a lot of, had a major impact, while strawberries, raspberries or blueberries had very little. Apples or pears only gave me a tiny peak, and were better for me than bananas. Drinking wine or beer had little effect, but orange juice gave me a massive spike, higher than Coca-Cola. These results are not likely to be the same for you and could not have been predicted accurately from the GI index of the foods (a measure of how much each food raises blood sugar), which are simply the average results of groups of people. In the same way we know that a size 7 shoe or a single car seat position is unlikely to fit everyone, I know now that I (along with virtually all of you) am not Mr Average.

Further direct evidence comes from the large DIETFITS study published in 2018 from my collaborator Christopher Gardner at Stanford University that had given 609 overweight or obese volunteers either a healthy low-fat or a healthy low-carbohydrate diet for a year.[7] No differences were seen between the groups and the major headline was 'it's a draw!' By reducing fat or carbohydrate intakes in each group by 30–40 per cent they on average had lost around 13 lbs or 6 kg in weight. But buried in the data, within each group, some people did much better than others, and some did much worse: while some lost as much as 60 lbs (27 kg) others actually gained 20 lbs (9 kg). For some people, even if they eat healthy non-processed foods, the random allocation of reduced carbohydrates or reduced fats just didn't work for them. With national guidelines insisting everyone should stick to the standard magic formula (such as low-fat foods), just imagine how many people are getting the wrong advice.

This research clearly shows that if you want to find the foods that work best with your metabolism, then you need to know your personal nutritional response – something that can't be predicted from simple online genetic tests. We all have our own personal tastes and preferences when it comes to food, so it makes intuitive sense to think that our personal metabolisms and responses to the foods we eat should be different too. But scientific research is only just catching up with this gut feeling, proving that everyone is unique and that there is no One True Diet that works for all.

Of course, there are healthy-eating messages that can apply to everyone – such as eating more fibre and plant-based foods and cutting down on sugar and ultra-processed products. But the take-home message is that there is no one right way to eat that works for everyone, whatever the glamorous Instagram gurus and government guidelines tell you.

2.
BREAKING THE FAST

*Myth: Breakfast is the most important
meal of the day*

'Go to work on an egg!', 'Breakfast like a king!' The idea that eating breakfast in the morning is the key to improving energy, concentration and mood throughout the day is a mantra ingrained in most of us from an early age. Over the last fifty years we have been bombarded with messages extolling the health benefits of various processed cereals, muesli and porridge oats. But what exactly is breakfast? The British full-fried extravaganza or the typical Italian equivalent, consisting only of a cappuccino and a cigarette? After all, a cappuccino, with its milk and sugar, contains all three macronutrients – carbohydrates, fats and protein – and will have the same effect on our metabolism as a bigger meal in 'breaking' the fast. What about a black espresso or tea without sugar, which can provide fibre and polyphenols but little in the way of energy? Many people who report that they skip breakfast every day will often have a milky tea or coffee first thing, meaning that they are still consuming something in the morning.

The lack of a good definition of breakfast highlights one reason why the research to date has generally been poor. Anglo-Saxon culture has suggested that breakfast has always been a part of our lives. Many contemporary food trends and fad diets, the Paleo Diet being just one example, stem from the desire to follow the natural example of our nomadic ancestors from millennia ago. This, however, is rarely discussed in relation to breakfast. When I stayed with the Hadza tribe in Tanzania, the last true hunter-gatherers remaining in East Africa, I noticed a definite lack of any breakfast routine, despite their regular sleeping patterns. They also have no singular word in their language to describe the concept of 'breakfast'. The men would wake up and usually leave on a hunting trip without eating, maybe grabbing some berries a few hours later en route. The women would stay close to the camp and would sometimes make some simple food like baobab porridge or eat some stored honey, but not usually earlier than 10am, meaning their diets included a fast of fourteen to fifteen hours, given the time they spent sleeping overnight. This contrasts with Western habits of fasting overnight for only eight to ten hours.

Although food historians disagree, breakfast has probably only become mainstream since the Victorian era – in past centuries we used simply to eat the leftovers from the night before. It also stands out as the one meal where people around the world now seem content to have the exact same meal for years on end, without getting bored. Indeed, people often feel lost without their familiar morning food routine, whether that consists of two slices of plain toast, a boiled egg, porridge or, further afield, dim sum, rotis or aloo saag. I remember struggling to adapt and eat my morning curry and vegetables when I worked as a medical student in an Asian hospital in Nairobi,

while Japanese and Korean breakfasts are often a total contrast to Western versions, usually involving rice, vegetables,
miso soup, pickles or spicy kimchi or fermented soybeans.

One reason why breakfast may be culturally and histori-
cally contingent rather than a universal human routine may
be related to the problem of storing food overnight and then
the hassle and time spent preparing it in the morning. This
meant that until the invention of modern fridges, you had to be
wealthy and have servants to enjoy breakfast. All this changed
with the invention of cheap processed food that could be stored
long term and eaten with minimal preparation. Kellogg's Corn
Flakes was the first major brand of processed cereal, invented
in 1894 initially as a health product and now eaten by the millions of bowls worldwide each day. Made of refined corn, they
have a high GI of 81, which is even higher than a potato at
78. After the nutritious, fat-containing parts of the grain are
removed, corn grits are superheated in pressure cookers for
several hours, then rolled flat and toasted. The result is mainly
toasted starch whose nutritional value is minimal, necessitat-
ing the addition of multiple chemicals and vitamins to fortify
it. The profit margins on processed breakfast cereals are over
40 per cent, allowing manufacturers to spend about 25 per
cent on advertising to keep children and young adults hooked,
as well as influence public and professional opinions on their
nutritional value. Because of the popularity – and profitabil-
ity – of breakfast cereal, there are now around 5,000 distinct
cereal brands in the US alone.

There are certain ideas about breakfast that are so widely
accepted, most people don't ever think to question them. For
example, that breakfast somehow 'kickstarts' our metabolism in the morning, allowing us to eat more efficiently later

in the day, as well as the idea that skipping breakfast makes you much hungrier later in the day so you overeat and put on weight. Despite a lack of evidence for these claims, they are presented as scientifically supported facts and clearly laid out in the UK in current NHS guidelines prepared by government employees from Public Health England, an expert scientific panel with input from the food industry. There are similar claims in the USDA diet guidelines for Americans and the current Australian Guidelines for Nutrition. These claims about the power of breakfast are mirrored in many other national guidelines, as well as by press and websites across the world. But what if we have been misled and this is just another diet myth?

A systematic review and meta-analysis of breakfast-skipping studies was finally published in the *BMJ* in 2019, on which I wrote an opinion piece.[1] They peer reviewed fifty-two studies and rejected most as they were of poor standard and not properly randomised (and so likely subject to bias), as well as four from low-income countries. Of the eleven randomised trials that passed muster, most came from the US and the UK, with one from Japan, and they varied widely in duration, from a day to six weeks, as well as in quality. Seven of the studies looked at changes in weight as well as change in energy usage as measured by metabolic rate. The conclusion of the meta-analysis is the same as previous reviews that were based on less evidence: namely, that there is no evidence to support the claim that skipping meals makes you put on weight or adversely reduces your resting metabolic rate.[2] Further, the data showed quite the opposite, with many studies providing evidence that skipping breakfast can actually be a useful strategy to reduce weight. Why has the field got it so wrong in the past and why did they not highlight the lack of good data earlier? There are a number of possible reasons, all linked to traditional beliefs surrounding nutrition and food.

Recent advice has been to eat little and often, i.e., 'grazing' rather than 'gorging' to avoid 'stress' on the body from having to digest large meals, which can produce large increases in insulin that could lead eventually to insulin resistance and diabetes. This advice is said to apply especially later in the day when glucose and insulin peaks are higher and metabolic rate lower. The rationale for this advice comes from multiple studies of small animals and a few short-term human studies. The pivotal study, the one that changed medical and nutritional thinking, was published over thirty years ago in the prestigious *New England Journal of Medicine*. In this study, men were given the same meals divided first for two weeks into seventeen mini-portions and, after a break, followed by the identical food given instead as only three meal portions. It was found that in the grazing group the insulin in blood decreased by 27 per cent, and a marker of stress (cortisol) decreased by 20 per cent. This all sounds impressive until you realise this famous study only involved seven people, and the results could be down to chance, and certainly should not be universalised to all humans and dietary requirements.[3]

The worry that skipping breakfast could lead to overeating later in the day is theoretically correct: people who miss breakfast do both usually eat more lunch and slightly reduce their physical activity. The body's metabolism is also initiated by a curious process known as diet-induced thermogenesis, whereby the food itself generates some heat in the body. But the key point is that these clever compensatory mechanisms, even when put together, are not nearly enough to make up for the weight-loss benefits entailed by the uneaten calories of a missed breakfast.

Certain misconceptions about breakfast, based on bad science, have become firmly embedded in nutritional dogma, even among experts. Nutritionists, doctors and food-industry professionals have, like the general public, been misled by headlines from several observational studies. These have shown that in any study of the general population, the people who skipped breakfast were more likely to be overweight. This is not the cause but merely a consequence of bias. Breakfast-skippers were more likely on average to have a lower income, be less educated and less healthy and have a generally poorer diet than breakfast eaters. All these social factors are independently associated with being overweight and the association has nothing to do with eating breakfast. Studies have shown that overweight people were also more likely to try and diet and, after a binge, be more likely to feel guilty and skip a meal.

Despite these obvious flaws, and the steady increase in contradictory evidence from randomised controlled trials, the idea that skipping meals was unhealthy has prevailed for decades and is still one of eight key healthy-diet NHS recommendations by Public Health England, as well as the current USDA diet guidelines for Americans and the Australian Guidelines for Nutrition.[4] Successful food industry conglomerates have a massive marketing budget and carry enormous clout with government officials, allowing them to influence public policy to the extent that verifiably false claims become government-sanctioned health advice. By examining what multi-billion-dollar industries like cereal manufacturing may stand to lose if more people begin skipping breakfast, it's easy to see why breakfast myths are so widespread and persistent.

So what are the benefits of skipping breakfast? Some of the positive effects of missing breakfast in the morning may be due to simply extending the time spent in the fasting state. Evidence is accumulating that time-restricted eating and increasing fasting intervals above twelve to fourteen hours can reduce insulin levels and help some people lose weight.[5] Some of these recent developments, which seem so counterintuitive to traditional thinking, make sense in the context of the importance of the gut microbiome. This community of 100 trillion gut microbes living mainly in our small intestine acts like an organ in our bodies, regulating our health and metabolism. The many microbes have a circadian rhythm similar to ours and vary widely in composition and function when in fasting and fed states.[6] Although this is a young field, some data suggests microbial communities suffer from long periods without food but could benefit from short periods of fasting, such as from skipping breakfast. After a gap of four to six hours without food, certain species start replicating and feed off the carbohydrates in the mucus of the gut lining, effectively tidying it up and making the gut barrier more efficient and healthier. Microbe communities, like us, may also need to rest and recuperate as part of a daily circadian rhythm – which could be important for our gut health.[7]

As well as false claims about reducing obesity, another common argument used in cereal marketing strategies is that breakfast is essential for aiding mental concentration in children. There are plenty of anecdotes circulating about children supposedly 'deprived' of breakfast running amok in the classroom or performing poorly because of low glucose levels. Again, the evidence is largely observational and likely biased in the same way as for adults.[8] Some independent reviews have looked at the many studies performed in this area and found them remarkably poor in quality. There were twenty-one short-term studies limited to just the effects of a single breakfast on attention later in the day. Only eight of these showed a positive influence, with the rest showing effects only in subgroups of undernourished male children. Similar results were seen with memory testing, and there was no consistency of results based on type of breakfast. These were artificial scenarios and are hard to extrapolate to the real world. In order to gain a more long-term picture of the effects of breakfast on academic performance, eleven studies were carried out on long-term school breakfast programmes. Seven of eight found no clear improvement in attention, and four out of five found no benefit for memory in the breakfast eaters. So there is no hard scientific evidence to favour forcing children to eat breakfast if they are otherwise well nourished. Some children and adolescents naturally don't feel hungry until later in the day.

Many people in developed countries around the world regularly skip breakfast. It is hard to get accurate numbers, although cereal and porridge manufacturers do sponsor surveys suggesting rates are increasing to 'dangerous levels' of nearly 50 per cent in countries such as the UK. As people get older they tend to skip breakfast less often and get into cultural routines. Many others, including myself, regularly enjoy breakfast. All of this is not to say that everyone overweight would benefit from skipping. Some of us are programmed to prefer eating food earlier in the day and others later, which may suit our unique personal metabolism and gut microbes.

Once again, then, there is no 'one size fits all' when it comes to the question of if and when you should eat breakfast. There is certainly no harm in skipping breakfast and I would recommend everyone try some experiments of their own to see how they feel without breakfast, both short term in mood and energy, and longer term in weight loss or gain after a month. If this feels too difficult, you could try occasionally skipping a meal. This is something I do to challenge my body's metabolism and also prolong my overnight fast to help my microbes. Breakfast may be the most important meal of the day – but only for some of us.

3.
CALORIE COUNTING DOESN'T ADD UP

*Myth: Calories accurately measure how
fattening a food is*

Calories in, calories out – this simple rubric defines the weight-loss strategy for hundreds of millions of people across the world. The diet industry is based upon this simple idea, but contemporary research is beginning to show that what we see as fundamental to a healthy lifestyle could be wrong – and even dangerous. While the central idea behind calorie-restricted diets is fairly self-evident – energy entering any living being equals the energy it puts out – most of us haven't a clue what a calorie actually is, despite it being displayed on every food label. Doctors like me were once taught about calories and kilojoules at medical school, but have long since forgotten any details. The general misconception is that they are a direct and precise measurement of how fattening a food is.

Antoine Lavoisier, a famous scientist living at the time of the French Revolution, was the first to understand that we 'burn' food as our source of energy. He measured the calorific units in food by inventing a bomb calorimeter, which was like a mini oven surrounded by water. This allowed food items to be burnt in order to measure their calorific value based on the heat that they each gave off to the surrounding water. He also fed unfortunate guinea pigs different diets and invented a similar device to put them in while alive, surrounded by ice, to observe the effects of how they converted food into heat energy. In the late nineteenth century, an American scientist named Wilbur Atwater dedicated his entire life to conducting experiments to determine the calorie count of over 4,000 food items. He not only measured the energy given off by the food, but would also feed them to volunteers and collect all the resulting heat, urine and stool samples produced. He would then carefully burn the urine and stool samples and work out how much energy they contained. He worked out that fat was roughly twice as energy dense as carbohydrates or protein, which was the beginning of the idea, now so intrinsic to our ideas about health and nutrition, that fat is particularly 'fattening'. His findings are still used today on food labels around the world and his work had a profound and lasting effect on the power and precision of the calorie as a unit to measure energy.

At first glance, this all seems to make perfect sense: we work out the calorific value of our food, then how many calories we need to eat to lose weight, and hey presto – we have a diet. It seems like a straightforward formula for weight loss, and it's easy to understand why the calorie has become a buzzword in the health industry. But although we can accurately measure the calorific value of a meal, the relationship between those calories and our body is much less straightforward. I discovered this for myself when, for a TV documentary, I spent twelve hours locked in a modern calorimeter in the University of Warwick. Under the supervision of Dr Tom Barber, a researcher into the human metabolism, the aim was to work out how much energy I produced. Most of our bodies burn through enough energy in twelve hours to light an 80W lightbulb. Entering the chamber, I felt as though I was descending through an airlock into a submarine, albeit one with big glass windows that would enable me to be observed. The container was surrounded not by water but by sensors, which would measure the rate I used up oxygen and expelled carbon dioxide. There was a basic bed, chair and desk, and some steps for me to exercise on. I spent the first few hours on the bed having a snooze so my resting metabolic rate could be measured. I then spent some time at the desk on my laptop writing notes for this book – no small feat in calorie terms, as the brain accounts for a third of our resting energy use. I then spent fifteen minutes exercising on the stairs to get my heart rate up and see how efficiently I could burn oxygen as fuel. Food was passed to me through an airlock and, in return, I thoughtfully produced some 'samples' for testing that I had eliminated in a small toilet behind a screen. After I was released, my results were calculated and the team were able to estimate my base metabolic rate – the amount of calories I would need to consume to maintain my current weight if I did no exercise whatsoever – as being approximately 1,600 calories per day.

According to international WHO guidelines, adult males are allowed 2,500 calories a day and females are allowed 2,000. This means that in order to prevent my body from storing extra calories as fat, I would have to be regularly burning off up to 900 calories through exercise during the sixteen hours I spend awake every day. I'm quite active, and I cycle to work each day, but an hour of cycling only theoretically burns off around 240 calories, leaving an enormous surplus to be made up by fidgeting, walking and thinking hard to burn off my fuel.

This is why I consider the idea of a universal recommended daily calorie intake to be misleading and, at worst, harmful. We can quantify to some extent what energy goes into our bodies, but not what gets burnt off. There are so many energy expenditure factors that vary widely between people, starting from your base metabolic rate, which can differ vastly depending on various factors including the total amount of muscle in your body and your fitness level. The base metabolic rate can vary by 25 per cent in normal, healthy people, from 1,450 to 1,900 calories. The amount of energy used up by exercise obviously also varies widely, as does the energy used up by small fidgeting movements throughout the day – a tranquil and restive couch potato could burn up to 10 per cent less energy than someone who finds it hard to sit still.[1] Finally, a small amount of energy is used up by just the physical act of eating and digesting food. By now, you should have started to form a picture of how difficult it is to estimate how many calories any particular person needs for their body to run efficiently and maintain a healthy weight.

So we can see that the science underpinning the idea of a daily recommended calorie intake is questionable at best, given the possibility for such sizeable variations from individual to individual in the amount of food necessary to keep them going. Other measures underpinning the supposed scientific objectivity of the calorie-controlled diet are also beginning to be called into question; namely, the accuracy of the original

Atwater experiments providing the calorie estimates for food labels. He did admirably with the science available to him at the time, and most of his estimates have held up with accuracies usually within 5 per cent, but the idea that we can measure the energy value of any food precisely is nonsense and claiming one dish with 312 calories is better than one with 329 is laughable.

As we gain a better understanding of the different components of food and how they interact together, some calorie-content estimates are emerging as inaccurate or outright wrong. Walnuts, for example, spent years with their calorie content inflated by 20 per cent on food packaging until it was discovered that much of the fat they contain is not released when we eat them. Almonds have been similarly over-estimated in calorie content by about 31 per cent.[2] Another good example of this is corn: how the body uses and stores the energy gained from a food as difficult to digest as corn on the cob is very different to how it uses energy from corn bread or from cornflakes processed by superheating, pressurising and roasting. Yet the simplistic calorie intake theory treats the energy gained from each as the same. We also now know that the way that food is cooked alters its structure and therefore how much energy it provides – so a raw beef-steak tartare will provide fewer calories than a bloody burger cooked rare, which will provide less than a well-done charred one. This is why the discovery of fire by our ancestors, and therefore of cooking, propelled human evolution exponentially forwards as the increased calorific intake allowed us to spend less time on eating and more on hunting and thinking. Eating cooked rather than raw food has also been shown to modify our gut microbes, making them evolve differently to other animals.[3]

To confuse things further, foods interact, and their calorific content varies when mixed, so the rate of energy released from a cheese sandwich, for example, may be different from the value of the bread and cheese measured separately. Even more importantly, the ultra-processed nature of modern food generally means that the complex structure of the plant and animal cells is destroyed, turning it into a nutritionally empty mush that our body can process abnormally rapidly. According to government figures the average person in the UK actually consumes slightly fewer calories per day than in 1976, although much more of that is now as ultra-processed food.[4] Some countries, such as the US, have introduced strict rules about displaying calorie counts in cafes and restaurants to help consumers pick healthier meals. Because these estimates depend on non-automated processes, and portion sizes vary enormously, studies have shown that the actual calorific content of a meal can deviate 200 per cent from the number on the menu and that restaurants nearly always underestimate the amount. There is little evidence that having these erroneous figures displayed on labels and menus helps with weight reduction.

The final underlying assumption behind calorie-restricted diets is that everyone burns the same fuel in exactly the same way and at exactly the same level of efficiency. My nutritionist colleague Sarah Berry pointed me to data that exposes this myth. When you look in detail at the results of the eighteen individuals in the almond study mentioned earlier, you see they vary threefold around this 'average' calorie estimate, with some people being good or bad metabolisers.[5] This means some people eating a handful of daily nuts would unknowingly consume around 700 calories more over a week than others. The calorie-burning assumption also ignores how and when you eat the calories. Studies in humans and mice are now showing that we put on less weight eating identical calories if they are consumed within an eight- to ten-hour window rather than grazing all day.[5] We already know that metabolic rates vary, but other factors, like the length of our intestines and how long food takes to pass through our digestive system, play major roles. Differences in our genes or the number of copies of genes we carry can make some of us extract more energy (in the form of sugar) from starchy carbohydrates like potatoes or pasta. Some people produce up to three times more starch-digesting enzymes (amylase) than other people, allowing them to break down starch and release more sugar far quicker. There is a simple experiment you can try to determine how well-adapted your digestive system is to processing starch: eat a basic wheat cracker and time how long it takes for you to taste sweetness. When we test our twins in the PREDICT study, we ask them to do it three times and take the average as it is not always clear-cut. We find about a quarter of participants can detect the change to sugar before thirty seconds, suggesting they are better adapted than most of us to eating starch, though we still don't know how this affects us individually.

Another factor that can lead to huge variation in our digestion is the make-up of our unique gut microbiomes. These microbes give us distinct chemical factories that provide us with different abilities to digest food and convert the contents into energy. Until recently, we didn't know how important the role of microbes was in human digestion and calorie loss, but several studies have now given powerful antibiotics to volunteers and measured the calories in their stools. Like any machine, our bodies are not 100 per cent efficient at processing energy, and we lose 2–9 per cent of our calories in our digestive waste, flushed down the toilet. These new studies showed that reducing our gut microbes (through the administration of antibiotics) could increase this wastage of calories by up to 9 per cent – a significant amount. This shows that variations between individuals in the amount and efficiency of our gut bacteria can have a big impact on how many calories our body absorbs and converts into energy. The message is clear: individual circumstances can vary to such an extent that the calorie as a measure of nutrition or weight loss is essentially useless.

We have all been brain-washed into seeing the all-powerful calorie as a weight-loss tool. This has blinded us to the fact that different foods of the same calorific value may have different metabolic effects on our body. For a long time, we only had a few lab animal studies to go on, some of which showed differences in weight gain for lab rats and monkeys fed different proportions of fat and carbohydrates. Recently, however, some studies on humans have been completed and these have clearly indicated that, although the number of calories gained from fat and carbohydrates may be equal, their effect on our bodies is not. One US study of 162 volunteers showed over twenty weeks that a high-fat diet, in comparison with a high-carbohydrate diet, appeared to increase the metabolic rate in volunteers significantly, so that they needed to eat 91 calories more per day to maintain their base weight.[6] It must be noted that this result was the average from a large sample size and that not all the participants responded in the same way, with some even seeing a drop in their metabolism. The US DIETFITS trial in 2018 showed that of 609 volunteers who spent twelve months on a calorie-controlled diet, some people lost much more weight when given extra carbohydrates and others when given extra fats, although the total energy intake was supposed to be identical.[7]

The other practical problem associated with the calorie is that even if you trust food manufacturers to measure and label their foods accurately using science that is often centuries out of date, it is still impossible to count your calorie intake accurately. Even trained dieticians with obsessive traits find it impossible to live a normal life and count to within about 10 per cent the calories they ingest. You would need to cook every meal by yourself from ingredients that you had weighed and measured with complete accuracy, taking into account the precise heat exposure of the food, and then eat an exact proportion of everything you cooked. Most diet plans depend on you counting calories or eating pre-cooked meals, and the errors of measurement are so great that even if you managed to follow the instructions you are unlikely to have a final calorie count that matches your recommended intake.

To understand our relationship with calories it may help to think of the petrol you put in your car. Imagine that before going on a weekly 200-mile trip in a car which, like your body, has no fuel gauge, you look up the average amount of fuel the average car would use during a trip of this length and then fill it up with that amount at the garage. You would then hope you were driving an average car at an average speed with average fuel efficiency and that you wouldn't run out of petrol. If you made a miscalculation and overfilled the car, and it kept the extra fuel in a reserve tank (like your fat cells), you might without knowing it slowly be making your car heavier and less efficient. Now imagine having a range of different fuels to choose from at the pumps, making it difficult to know which of them best suited your engine.

The biggest problem with the calorie is not the measurement itself – which does serve some crude purpose in allowing us to distinguish the relative energy gained from eating celery vs eating potatoes – but in giving us a false sense of security and precision. The use of the calorie has been great news for the food industry, sending sales of 'low-calorie' foods and snacks sky-rocketing and keeping marketing departments busy, while health regulators have been able to show their governments quantifiably that they are doing something positive. But the calorie has been a disaster for the average consumer. We have been fooled into thinking that food is something that can be spoon-fed in precise quantities, and conned into eating low-calorie diet foods that fail to satisfy our hunger and made by diluting real food with nutritionally empty chemicals. This obsession is still being encouraged by the UK government who have made calorie counts on menus in large restaurants and cafés compulsory from 2022 – a change that will likely only accelerate the nation's unhealthy relationship with food. We're not cars with a fuel gauge; we are far more complex and intricate, and rather than basing our decisions about what to eat on a universal, arbitrary and often inaccurate number, we need to learn to understand our individual bodies and what they need.

4.
THE BIG FAT DEBATE

Myth: Saturated fat is a major cause of heart disease

About twenty years ago I was sufficiently convinced by the weight of evidence to switch from eating butter to a low-fat spread made with what appeared to be Italian olive oil. My wife, who is also a doctor but raised in a different food culture (French–Belgian), had no intention of changing the diet habits of her ancestors because of a few medical papers. We agreed to differ and kept both butter and margarine in the fridge. About five years ago, a few dissenting voices around the world began to question the wisdom of telling people to avoid butter. But as often happens, there was a backlash. Over a hundred angry academics signed a letter to the editors of the *British Medical Journal* in 2018 criticising their supposed pro-butter stance and for allowing publication of a 'biased' editorial comment that claimed the saturated-fat heart story was grossly exaggerated and we had been misled. They were also upset about some extreme claims in the editorial about the lack of benefit of statins. A rebuttal followed.[1] What should have been a sensible scientific debate was then inflamed by newspaper articles accusing the authors of 'religious fundamentalism'. This then degenerated into an unpleasant personal slanging match over bias, integrity and beliefs.

There is certainly a touch of religious wars about these debates, and much of this is because of who controls the sacred nutritional guidelines. The current ones in the US and UK are similar and both have areas of common consensus that few disagree with, such as eating fewer calories, more plants and vegetables, less processed food, and drinking fewer sugary drinks.[2] Where there is less consensus on the evidence is on whether we should cut down on saturated fat and, if so, what to replace it with.

The diet–heart hypothesis is at the centre of this issue and is far from simple. The original idea that started in the 1960s was that cholesterol in food was to blame for heart disease by increasing levels of cholesterol in the blood, which caused furring up of the arteries. This has been disproven, and no serious scientist believes it today. It came from severely flawed observational studies, and we now know that our liver naturally produces most of the cholesterol in our bodies and that cholesterol in food doesn't alter its levels in the blood to any extent. Many foods we now think of as healthy contain large amounts of cholesterol, essential for the health of our cell walls and a number of key vitamins. All animals and animal products contain cholesterol – red and white meat, oily fish, eggs and yoghurt all contain high amounts. Some of these, such as eggs, may continue to be linked in headline-attracting observational studies to heart disease, but others, like fatty fish, are paradoxically billed as protective. Moreover, the evidence is weak and the risk often trivial.[3] Meanwhile, food marketing continues to promote certain products misleadingly as choles-terol lowering. For example, plant (or phyto) sterols are added to many foods or oatmeal, yet the effects on blood are mini-mal at normal doses and there is no good evidence it reduces the likelihood of heart disease. The original 'cholesterol was bad' hypothesis was replaced around 1980 with the 'all fats are bad' hypothesis. This came largely from the fact that fat is more calorie dense than the same weight of carbohydrates or protein, along with the idea that layers of fat get deposited in your blood vessels, which leads to heart attacks. The food industry was – and still is – happy to promote low-fat foods using highly processed chemicals and sugar to replace expen-sive dairy products, and labelling it all as healthy.

Fat, however, is not a single entity – it is a vague umbrella term for anything made up of building blocks of three fatty acids joined together (to form triglycerides) that make up 98 per cent of the fat in our diet. Each fatty acid can vary enor-mously in length and number of connecting bonds that alter its properties, making them saturated or unsaturated and more or less solid. These saturated or unsaturated fats may be identical in plants and animals, but always exist as a mixture in the same food. There is more saturated fat, for example, in a tablespoon of olive oil than a lamb chop, but both foods contain many other fat subtypes (such as mono- and poly unsaturated fats) that have different effects on the body. We construct most fats in our liver, but are unable to make from scratch essential poly-unsaturated fats such as omega-3, which we get from our diet.

When the different types of fat – the good, bad and the ugly – are combined, they are called total fat. By the 2000s, clinical trials were proving that reducing total fat in diets has no effect on health, leading to the slow demise of the 'total fat is bad' hypothesis in the US, though not in the UK. However, unwilling to give up on the idea that fatty foods were harmful, most nutrition scientists backed the saturated-fat hypothesis. This is based on largely observational (and some genetic) data from Western populations that the levels of a small particle containing cholesterol in blood called LDL (low-density lipo-protein) is correlated with heart disease and that these levels are slightly increased proportional to bigger, healthy particles as the amount of dietary saturated fat increases. This data did not distinguish the food sources of the fat and was thought to be solid until quite recently, when large observational stud-ies from poorer countries in different environments provided the opposite result. For over seven years, 135,000 people from eighteen countries were followed as part of the PURE study and the results showed that people eating dairy and higher saturated fats were associated with lower mortality than those eating more carbohydrates.[4]

Importantly, despite several attempts, no study has suc-cessfully shown that switching from a normal or high-fat diet to a low-fat or low-saturated-fat diet can reduce heart disease or mortality. Large-scale trials like the PREDIMED study, which divided 7,000 Spaniards into a low- and high-fat group, have actually shown the opposite effects, with heart problems and deaths reduced by a third in the higher-fat diet group – though this was mainly via total fat rather than purely saturated fat.[5] Despite the new studies and lack of convincing evidence, guidelines continue with the reductionist idea that foods containing saturated fats are bad for all of us. The pos-sibility has been largely ignored that the observational stud-ies coming from the US showing saturated-fat intakes were associated with heart disease may just be general markers of unhealthy lifestyles. Eating fried food regularly in the UK and the US is associated with being unhealthy, whereas in Italy and Spain eating deep-fried seafood with plenty of saturated fats is common, but often accompanied by salads. Chemicals in the salads may interact with the fats to produce other heart-protective chemicals (called resolvins).[6]

Another food that has swung from hero to villain and back again because of its saturated-fat content is the egg. Eggs are about 11 per cent fat, mostly cholesterol and some saturated fat. Studies based purely on observing large populations over time have not found any big problems of eating up to one egg per day, with some studies outside the US showing protective effects.[7] Most countries have discouraged eating them regularly to be consistent with their cholesterol diet policy. The US recently reversed its advice and said they were now safe to eat in moderation, but as they said other fatty foods like meat should be reduced, this inconsistency may have been due to lobbying from the American egg board.

Another problem if you label some groups of fats as unhealthy is deciding what practical advice you give people to make changes when many foods considered healthy contain some of these fats. The current recommendation is that foods containing saturated fats should be replaced with either starchy carbohydrate foods or unsaturated fats. This means, for example, that we should swap butter for low-fat spreads (the rebranded name for margarine). The desperate wish to have a simple message applicable to everybody – such as 'reduce all saturated fats' – creates the problem. This reductionist approach ignores the complexity and quality of foods, dietary patterns and individual food choices – and it totally neglects individual variation.[8] Research thinking is shifting rapidly from food not merely being about macronutrients and calories, but about the hundreds of chemicals in fatty foods that interact with each other and our trillions of gut microbes – which are unique to each of us.

Even if you ignored our individual differences, combining data from multiple studies of 635,000 people in fifteen countries by independent experts shows that eating butter, which contains large amounts of saturated fat, does not on average cause us any harm.[9] Strangely, there is no similar long-term data on the health effects of eating modern spreads, probably because we were told they contain 'healthy' fats. This is despite the fact that the early margarines we were encouraged to eat in the 1980s and 1990s were very harmful for most of us. These early margarines contained trans fats that were chemically engineered by the food industry to allow the fat to be solid at room temperature. Our bodies couldn't deal with these artificial creations and they increased the risk of heart disease threefold, killing an estimated 250,000 Americans each year. The food lobby delayed them from being phased out from most Western countries for about a decade before the tide of opinion changed.

Unfortunately, the lessons from the past haven't been learnt. Telling people to eat low-saturated-fat spreads instead of dairy will force people to consume cheap highly processed items with multiple additives and new industrial fats that we know little about. Fats created by a complex process called inter-esterification have now replaced trans fats in biscuits, snacks and spreads in most countries. Inter-esterification is a process of shuffling the fatty acids around within the fat molecule, allowing the food industry to blend highly saturated fats (such as stearic and palmitic acid) with an unsaturated oil to produce a fat with suitable melt profiles for many products.

These are generally considered safe, but as the industry continues to tinker with the ideal combinations, they are slowly being introduced into many of our foods without any proper long-term human testing.

Some high-quality vegetable-based spreads with minimal processing and plenty of polyunsaturated fats may be healthy for some people, but globally people are confused and suspicious of the messages and often ignore official guidelines. Unilever, the market leader, recently sold their spread business because of market trends, and sales of 'natural' butter are increasing at the expense of 'artificial' low-fat spreads. Even doctors, who should understand more about fats and diet than the average person, don't follow the guidelines. After I wrote an opinion piece in the *BMJ* recently, around 2,000 GPs replied to an online poll as to whether they followed the UK guidelines to eat low-saturated-fat margarine rather than butter, and 83 per cent said no – they ignored the advice.

Two things need to change if we are to improve our health and diets. First, demonising one major food group (or type of dietary fat) is a mistake. Foods contain a wide range of saturated, mono- and polyunsaturated fatty acids in varying proportions, and the different fatty acids never exist in isolation. This means that fats in the food we eat can have contrasting health effects, depending on how the subsequent digested fat particles that float around in our blood vessels actually affect us. Second, we need to say farewell, once and for all, to the idea of the 'standard human'. We are more individual in our food responses than we think. The 2018 DIETFITS trial of 609 overweight adults from colleagues at Stanford in California found an equal number of people lost weight on a low-fat high-carb diet as on a low-carb high-fat diet when followed for a year.[10] Within the high-fat group many people saw big weight losses and others failed to lose anything. Assuming that fat is bad for everyone is clearly wrong and one size of dietary fat advice is highly unlikely to fit all.

It is dangerous to vilify doctors and scientists who criticise guidelines or question population approaches to diet as religious fanatics. In the last year alone, we have seen major clinical beliefs that have been around for years debunked by new data, such as aspirin for primary heart prevention, vitamin-D supplementation for preventing fracture, low-salt diets for preventing heart failure, and omega-3 supplements helping diabetes. All these have now been shown not to work. We need critics and open debate more than we need outdated and inflexible guidelines.

As for my personal choice, I realised about seven years ago that my wife was, as usual, right. She had never overreacted to scaremongering about butter, and I threw away my pot of polyunsaturated vegetable spread with a tiny amount of low-grade olive oil, preservatives and yellow food colouring and pictures of happy Italian peasants and returned to good old straightforward butter. But the choice is yours.

5.
THE SUPPLEMENTS
REALLY DON'T WORK

*Myth: Taking vitamin supplements improves
our health and prevents disease*

We've become nations of pill poppers. With half of all Americans and Brits taking a daily supplement and around a billion regular users worldwide, we're evidently still obsessed with vitamins a century after they were discovered. We are told they work for everything from treating cancer to preventing hair loss, curing hangovers to boosting energy levels. By 2025, we will spend $193 billion a year on these products globally, making the companies that produce them as rich as the pharma giants. The wealthy can now even get their daily dose of goodness through IV vitamin infusions.

We have fallen in love with the idea that adding vitamin D, calcium, folic acid, B vitamins and iron to an ever-greater range of foods, including milk, breakfast cereals, bread and many ultra-processed foods, makes them 'nutritious'. This fortification allows public health departments to say they are improving the health of the nation 'naturally' and cheaply without appearing 'medicalised'. But we wildly overestimate the benefits of supplements and underestimate the risks. Virtually none has been proven to work, and the evidence increasingly points towards their doing more harm than good.

The idea that we need extra vitamin and nutrient supplementation in our food originates in the widespread nutritional deficiency of the 1930s, and it has persisted, despite over-nutrition with junk food being the current problem. This mindset that vitamins make food 'healthier' lacks any evidence from studies, but it has allowed food companies to exploit us into thinking that highly processed foods with added vitamins and minerals are somehow good for us. For example, companies fortify sugary cereals such as Frosties (Frosted Flakes), as this enables them to legally make health claims such as 'good source of vitamin D', camouflaging the fact that a small bowl contains over half of a child's recommended daily sugar intake.

A lot of the public confusion and hype surrounding supplements stems from clever marketing campaigns, anecdotal reports of supplements curing diseases, or even old wives' tales. For example, the myth that taking vitamin C will boost your immune system probably comes from Nobel Prize winner Linus Pauling, who in the early 1960s hypothesised that taking vitamin C could prevent the common cold, despite many reliable studies disproving his theory. A few studies have shown that if taken with zinc supplements, vitamin C might reduce cold symptoms by an average of about six to twelve hours. However, you can probably get similar benefits from drinking a glass of orange juice or eating a kiwi fruit, it's just that no one has paid for proper studies to be done.[1]

Eating a healthy, balanced diet, one that includes lots of brightly coloured fruits and vegetables along with some fish and small amounts of dairy and high-quality meat, plus daily exposure to the sun, should provide enough vitamins and minerals for 99 per cent of us. Moreover, our gut microbes produce some vitamins, such as the vitamin B family, folic acid and vitamin K. The UK's fruit and vegetable intake has barely changed over the past decade, with all age and sex groups consuming below the five-a-day recommendation, while 90 per cent of Americans fail to meet the federal guidelines for fruit and vegetables of four to five cups per day, despite strong evidence that these foods reduce our risk of disease. People think that because studies have shown favourable effects of eating foods like fruits, vegetables and oily fish on our health, consuming a few of the chemical components found within these foods as supplements will bring about the same health benefits. We know from large clinical studies that this isn't true.

I confess that I used to take supplements, including vitamin D and omega-3 fish oil, but six years ago I changed my mind. In the course of writing my books and publishing over thirty research articles on vitamin D and calcium, the properly conducted and unbiased projects that I studied proved to me that supplements not only don't work, but in many cases they could actually be harmful. The healthy supplement hype is fuelled by governments and health systems, despite increasing evidence to the contrary. We seem to have forgotten that vitamins are just chemicals under a different name. The British government is keen on fortifying many foods and recently stated that the evidence of vitamin D being good for our health was so overwhelmingly strong, that all 60 million people should take a daily supplement for half of the year.

In the US, unlike drugs, dietary supplements are poorly regulated by the Food and Drug Administration (FDA). This means that the thousands of dietary supplements that fill US pharmacy shelves are not evaluated for safety or efficacy, or even their true contents. In 1991 an act was proposed to regulate this growing problem, but the industry successfully lobbied Congress to pass the controversial 1994 Diet Supplements Act using a series of adverts about personal freedom. This extraordinary act means the FDA cannot question any supplement company's data, contents or claims without doing their own expensive research studies on the 85,000 different supplements on sale. This has created a 'Wild West' atmosphere where anything goes. Even in Europe and Australasia, no safety checks are needed, nor even warnings on the label, including on St John's wort – a supplement that interferes with many common medicines.

Companies around the world still have free rein to make exaggerated or misleading claims. What started as a cottage industry is now a global business. The regulation of claims in Europe may be slightly stricter, but companies can still, for example, assert that their products 'boost energy' if they contain some approved ingredients, such as a trace of zinc, even if the study from which this assertion came is forty years old. As vitamins are tiny, most tablets or capsules need to be padded out by bulking agents, preservatives and multiple other minor chemicals or waste products that are never checked. Multivitamins often have extra hidden ingredients and crushed-up Viagra or anabolic steroids have been found added to some 'multivitamin tonics'. Studies observing over half a million people taking these unregulated multivitamins have shown they are more likely to develop cancer or heart disease.[2]

Vitamin D, aka the sunshine vitamin, is the poster boy of the supplement world and is commonly believed to have the greatest evidence supporting its use. I studied it as an academic for twenty-five years, led the group that found the genes influencing it and wrote over twenty scientific articles about it, including a placebo-controlled clinical trial of supplements in normal menopausal women. I believed it could prevent disease, and that we should take more of it. Vitamin D was originally given to Victorian children living in urban poverty to cure rickets, and it's now routinely given to millions to prevent fractures. I certainly recommended it to most of my patients with bone or joint problems. Its benefits apparently extended way beyond bone, and hundreds of observational studies have shown low levels in nearly all common diseases, including autoimmune disease, heart disease, depression and cancer.[3] I have now, however, changed my tune and believe for most people (except the bedbound and a few other rare exceptions, like MS sufferers) it does not work, and the risks outweigh any benefits.

These observational studies are all skewed, because the disease usually causes the deficiency rather than the other way around and people have confused low blood levels with actual disease. More importantly, summaries of the high-quality randomised control trials fail to show that supplements work. Data from the largest ever clinical study on the benefits of vitamin D in preventing fractures was reported recently, with over 500,000 people and around 188,000 fractures from twenty-three cohorts from many countries using vitamin D genes as surrogates of blood levels.[4] There was no effect whatsoever of either vitamin D or milk consumption, and therefore calcium, on the risk of fracture.

Vitamin D isn't actually a vitamin, since our body can make it naturally from chemicals in the skin on exposure to sunlight. It should be called 'steroid hormone D', although presumably this would make it much less popular. It is fat-soluble, meaning that like vitamins A, E and K, toxic levels can build up in the body as it is stored in fat tissue. While recommendations for supplements are usually modest doses, these will inevitably be overdone by many people, including those who buy high-dose supplements on the internet. Although vitamin D toxicity is rare and causes high levels of calcium in the blood, it has serious effects on the heart, kidney and brain, which can last for months. Incidences of toxicity have been increasing over the last few years, and look set to continue, due to internet sales and people eating fortified foods.[5] Despite popular belief and marketing, you can get enough vitamin D from fifteen minutes of daily sunlight exposure, or by eating a fillet of oily fish such as salmon, or a handful of vitamin D-rich mushrooms.

In the US, and increasingly the UK and Australia, milk, cheese, yoghurt, breakfast cereals, fruit juice and even water are routinely fortified with vitamin D, iron, calcium and folate. They are hard to avoid and overdosing is becoming increasingly common. Iron overdose is becoming a problem in the US, where it is added routinely to many foods, including pasta and cereals, often as low-quality products like elemental iron (iron filings). Folate supplementation in women trying to conceive was one of the big vitamin success stories of recent years, reducing birth defects by about 70 per cent, and this led many countries to add it to everyone's food. France has resisted the trend, deciding there could be risks, and, to this day, won't allow indiscriminate food supplementation. There is growing evidence that the French may be right and that high levels of folate can lead to health problems and increased cancer risk. Many vitamins have a slender therapeutic range of benefit, which is easy to regulate in real food, but difficult when added as a chemical. People assume that the production of vitamins is a natural organic-focused cottage-industry, but they are mostly manufactured synthetically in enormous factories, often in China. The top hundred vitamin brands are owned by fourteen global corporations, including Nestlé, P&G and Bayer.

Overuse of vitamin D supplements has been linked in several trials to weakened bone density, as well as increased falls and fractures.[6] Calcium supplements, meanwhile, have been linked in trials and genetic studies to modestly increased risk of heart disease and strokes, potentially due to the calcium clogging and damaging key arteries.[7] Our bodies can't deal with a large dumping of a chemical supplement in our intestines in the way that they can process and absorb them from natural food sources.

Arguably, some of the biggest current fads are protein supplements and high-strength water-soluble vitamins, both of which when consumed above our nutritional requirements are excreted out of the body, meaning the extra doses generally end up in the toilet. Protein supplements are the heavyweight in the $16-billion sports nutrition world and they're reportedly used by up to 40 per cent of Americans and 25 per cent of Brits in 2016. Far from being protein deficient, most healthy people in Western countries exceed the daily recommended protein requirements, yet marketing tells us otherwise. The food industry have jumped on the bandwagon, adding a few extra grams of protein to chocolate or granola bars in order to proclaim that their calorie-laden products that used to be high energy are now 'high protein' and the perfect snack to slip into your gym bag.

Everyone knows protein supplements help to build big biceps and bulk up the body, and this is the basis of selling protein supplements as powders or drinks at a hundredfold mark-ups. But while strength athletes have higher protein requirements than your average couch potato, this is a tiny difference, only around 50 g per day that can easily be met through eating a single chicken breast or a can of baked beans. There is also no difference between plant and animal protein in building muscles, so steak and egg diets are unnecessary. Some studies, including many small ones sponsored by the food and diet industry, have suggested protein snacks or drinks can help muscle recovery within a forty-five-minute window of time after stopping. Many better studies now show no advantages compared to eating protein before exercise.[8] You can therefore avoid the expensive supplement and get the same benefit from a glass of milk and a handful of nuts on your way to or from the gym.

While taking high levels of protein is no longer considered harmful to our kidneys, many of the popular supplement brands contain a long list of additives, chemicals and flavourings that have not been tested properly.[9] If you normally choose a whey- or soy-based protein powder after your workout, you'd be far better off eating these high-protein foods in their natural forms as a milkshake or a high-protein stir fry at home. Unless you're a professional athlete, if you're eating a few protein-rich foods daily, you'll almost certainly be meeting your increased protein requirements.

Why do so many people think that high-dose supplements are superior to real food? Take the humble tomato, for example. It's naturally rich in lycopene, a powerful antioxidant which has been linked with reductions in the risk of heart disease, yet many people buy high-dose lycopene supplements online instead. A review of over thirty studies suggests these people might be wasting their money. The real tomato generally performed better, probably because of the hundreds of other chemicals it contains. There are similar studies with other vegetables like broccoli.[10]

Some supplements have been shown in large trials to actually *increase* the risk of disease. One large trial in 2014, for example, suggested that vitamin E and selenium supplements did exactly this to prostate cancer rates.[11] For years, I – along with many others – believed that fish oil capsules containing omega-3 fatty acids are good for us, as they are marketed as a cure-all for arthritis, heart disease and dementia. This has been further perpetuated by organisations such as the American Heart Association, who suggest that people with coronary artery disease might like to top up their omega-3 intake with supplements, a decision perhaps related to a combination of naivety, sponsorship, advertising and a wish to empower individual sufferers with simple solutions. Americans spend over a billion dollars a year on fish oil supplements. A recent review that combined seventy-nine randomised trials involving 112,000 people concluded, however, that taking long-chain omega-3 (fish oil, EPA or DHA) supplements neither benefits heart health nor reduces risk of stroke or death from any cause.[12] A large 2019 trial of 25,000 Americans found no benefits of fish oil on preventing heart disease or cancer.[13] Other substantial trials have now shown that fish oils fail to prevent blindness, Alzheimer's or prostate cancer. Spin-offs such as omega-9 supplements are another massive fraud, selling us non-essential fats that are in almost every food already. Our ignorance makes us vulnerable.

Despite what the government, public health bodies and food companies tell us, healthy people don't need supplements. Instead, we should simply eat a diverse range of fresh foods and get a few minutes of sunshine a day. For 99 per cent of people, this will provide all the healthy vitamins and minerals you will ever need.

If after reading this, you do continue to take supplements, make sure you know exactly what and how much you're taking, as too high a dose of these chemicals could have harmful effects on your health. Remember that virtually no vitamin or mineral supplement has been shown to have any benefit in proper randomised trials in normal people, and increasingly they are being shown to risk causing harm.[14] If these chemicals were not protected by being called supplements, they would be banned. More and more processed foods are being fortified each year by food manufacturers simply because they see financial opportunities to combine cheap ingredients with health claims.

We fall for these fads because we want a quick fix and a miracle health boost, and a daily vitamin and mineral supplement might seem like the perfect solution. Everyone likes to feel that 'they are doing something good for themselves'. But popping pills won't outdo a bad diet and the science behind supplements simply doesn't stack up.

6.
THE BITTERSWEET
HIDDEN AGENDA

Myth: Sugar-free foods and drinks are
a safe way to lose weight

As sugar taxes bite around the world and public perception that highly sugary drinks are unhealthy takes hold, the switch to and acceptance of artificially sweetened foods and drinks is increasing. Government recommendations in most countries favour drinking water, but switching to artificially sweetened beverages (called ASBs in the trade) is considered a suitable alternative. The fact that these drinks contain near-zero calories means that they are often called 'diet sodas' and health agencies are happy to endorse them, implying that they are safe and will help weight loss. Similarly, with official emphasis focusing on reducing calories at all cost, consumers are being steered towards low-calorie ultra-processed foods or desserts packed with artificial sweet chemicals rather than sugar.

The average can of Coke, Pepsi, Sprite or Fanta has around 8–12 teaspoons of sugar depending on the country, providing roughly 140 'empty' calories. For people consuming two cans a day, this is over 10 per cent of their total daily recommended calorie (energy) intake. In 2013, the US produced around 160 litres of sodas per person, which is around one can per person per day. The UK produces around half that amount per person, which still leaves over 81 litres per citizen. Given the huge amounts of sugar and calories consumed, it is quite logical that switching to zero-calorie sodas should lead to weight loss and fewer health problems. This is one reason so many people (one in three Britons and one in four Americans) have switched.

But are ASB sodas really that much better than regular sugar soda? A meta-analysis in 2019 summarising fifty-six studies, of which seventeen were randomised controlled trials, found many were too small and of poor quality. But when analysed together, adjusting them for size and quality, the study showed no clear benefit on weight loss of taking regular ASBs compared to regular sugar sodas.[1] So what is going on? How can these inert substances with near-zero calories be any better than drinks with added sugar, apart from reducing the amount of tooth decay?

The world's most common sweeteners are sucralose and aspartame, both of which are often combined with another chemical, AceK. These chemicals are added to over a third of all soft drinks sold, plus many other processed products like yoghurts or low-sugar 'healthy' foods, like chewing gum, biscuits, vitamins, medications and toothpaste. Artificial sweeteners allow for cheaper and more efficient production of a wide range of highly processed foods, help to extend the product's shelf life and make much of our food sweeter. Sucralose or aspartame are 200–600 times sweeter than sugar, and are

increasingly being used in combination with real sugar in processed foods and sweets.[2] Children are constantly exposed
to artificial sweeteners, which may drive them towards consuming ever sweeter foods, a habit that can persist as a mild
addiction in adulthood. The industry now appears to be using
artificial sweeteners not just to reduce calories but to stra-
tegically modify consumer behaviour. Cigarettes, cigarillos and
new smokeless tobacco products are more popular and addictive when artificial sweeteners are added. What works for the
tobacco industry, which has decades of practice in addiction,
also works for food. Targeting your future customer is a crucial
part of marketing: in this case, children.

Since the 1970s, consumers have been worried that artificial sweeteners cause cancers. This started with stories of early
chemical sweeteners such as saccharin, which was made from
coal tar and cyclamate over a hundred years ago, causing bladder
cancer in rats. But these studies were never fully replicated or
shown to be a real problem in humans. Rats also had to ingest an
equivalent of several hundred cans of diet drinks per day to show
any effects. Initially, the drinks manufacturers were annoyed by
this focus, but quickly found alternative solutions, which they
could prove were 'cancer free'. There was never any good evidence linking cancer and sweeteners, but while these fears have
continued to worry consumers, they have turned out to be a useful distraction for the industry. The safety regulators focused on
whether any new sweetener gave rodents rare cancers, which
was usually disproved. This meant that the link between these
artificial chemicals and other major health problems, such as
diabetes or obesity, were overlooked or downplayed.

This distraction and smokescreen was encouraged by the
powerful drinks industry which distorted and manipulated the

views of nutrition researchers and experts. Leaked emails have recently uncovered the extent of the money used to influence opinion, revealing that the US HQ of Coca-Cola had spent $140 million between 2010 and 2017 solely on giving research grants to academic scientists, and Coca-Cola and Pepsico gave even more to ninety-five US health authorities, of which sixteen were medical organisations.[3] This ensured academics were kept distracted and busy, writing papers showing that their sugar or artificially sweetened products were safe, and that lack of sports and exercise was the main cause of obesity, not harmless sweet drinks. An investigation in the UK by the *BMJ* in 2015 uncovered a similar web of funding key decision makers in nutrition by the sugar industry.[4] This practice of sponsored research is commonplace in multiple countries, and the amount of money channelled via these 'grants' is enormous compared to the tiny amounts most nutrition departments receive from standard government or charity grants.

It's not surprising, then, that many academics take part in these small studies to keep their careers going. Drinks industry-funded academic studies of foods or drinks are twenty times more likely to find a favourable outcome for the sponsor than independent ones. The industry also supports data summaries by 'independent' organisations who will provide the desired result for the right fee. At least 30 per cent of over 400 or so studies on artificial sweeteners are funded by the drinks industry, and although there are exceptions, these studies tend to be of insubstantial size and are mainly in rodents, adding to confusion.

This is just the subtle side of the influence. In Mexico and South America, the largest and most lucrative market for the drinks industry, the stakes are even higher. Anti-sugar campaigners have been physically intimidated by hired thugs, and politicians have been persuaded to drop legislation at the last moment with the promise of funds for their next campaign. This undue influence of the drinks industry, and sometimes entire countries, on researchers and regulators and possibly journalists, explains why it has taken so long for us to realise that ASBs offer no benefit in weight loss over regular fizzy drinks or sodas and why no one objects to them still being called diet drinks.

So why don't ASBs work as they should for weight loss if they have negligible calories? The drinks industry tells us that these synthetic molecules tickle our taste receptors, replicating the effects of regular sugar with the added bonuses that they neither contain any calories nor alter our metabolism, passing instead through the body like stealthy ninjas. But what if this is just a fairy tale? During several of my diet experiments where I was wearing a continuous glucose monitor, I made myself a drink with several sachets of sickly-sweet sucralose and water, which I downed like a syrupy shot. On two out of three occasions I tested my blood sugar levels; they surged by 30–40 per cent after thirty minutes and then reduced back to normal. During one of these experiments, I was inside the tightly controlled environment of a metabolic chamber and the brief rise in blood glucose thirty minutes later that was not supposed to happen was clear. It was obvious that sucralose was affecting my gut as well as fooling my brain and taste receptors.

Two studies shed light on possible mechanisms for this: one investigated the effects on the brain and the other looked at the gut. The first, conducted in 2017, used fifteen volunteers all of average weight who were fed five different drinks over a few days while lying in special head scanners that measure brain activity by lighting up active areas. These functional brain scanners were used a bit like lie detectors to avoid possible bias. The five drinks had different chemical mixes for sweetness and calories but there was no way for the volunteer to tell them apart.[5] The researchers found that the reward centres of the brain lit up more when given a sweet drink containing the sweetener sucralose (and so with no or few calories) than a drink containing real calories. They speculated that the mismatch between the perception of sweetness and lack of calories messes with our brain, sending out the wrong metabolic messages to our body. When our brain finds that it has been conned and the expected energy never arrives, it tries to regain this energy by storing fat or reducing activity. The reality is that we don't know for sure, and the effect of these chemicals is going to be much more complex as they all work in different ways.[6]

The second key study was from Israel in 2014 and explored whether gut microbes might be involved in the potential side effects of obesity and diabetes. A range of ASBs were first tested for their effects on the gut microbes of mice. It turned out all the common sweeteners (sucralose, aspartame and saccharin) altered the composition of the microbes of the mice leading to abnormal blood sugar levels. When these microbes were transplanted from sweetener-fed mice to sterile, germ-free animals, these altered microbes increased the blood sugar levels of the new hosts. When antibiotics were added to kill many of the gut microbes, the abnormally high blood sugar response was abolished, showing that the microbes had been crucial to the spiking effect. Seven human volunteers were then given saccharin. In four, high glucose peaks were caused while in three, there was no response. The microbes of the human responders were then transplanted into sterile mice and similar blood effects were seen, showing these changes were causal, not consequence.[7] There have now been multiple studies with similar conclusions, with the data consistent for saccharin, sucralose and the sugar alcohol xylitol – although most of these were performed on mice, with some doubts about appropriate dosing and quality.[8] Pigs are more similar to humans and breeders use artificial sweeteners like saccharin to alter the gut microbiome of piglets to help rapid growth.[9] While it makes sense for farmers to fatten up young pigs rapidly, the same is not true for modern humans. The drinks industry is understandably very concerned about this new work on ASBs and the microbiome and the negative direction the science and publicity is taking.

Although we have limited human data so far, we are collecting more data as part of our PREDICT study by giving our twins sachets of sucralose and aspartame to test.[10] It is looking like our responses to these chemicals are highly individual, probably because of our different microbes. So far, around one in six people have clear but unexplained glucose peaks after testing, while most have more subtle or no changes. I found no clear glucose signal when I took the aspartame or AceK sweeteners, but these may have other effects on my body I couldn't measure directly.

These chemicals, which can only be made in labs, were mostly found by chance by some chemist licking their fingers and getting a sweet surprise. They also differ widely, with some like sucralose not being absorbed into the blood and staying in the gut and others like AceK being rapidly absorbed into the bloodstream. A 2019 study of 154 people over twelve weeks found that the top sweeteners all had different effects on body weight, with saccharin being close in effect to sugar, aspartame producing slight weight gain and sucralose proving slightly beneficial, but individuals are likely to react to each chemical noticeably differently.[11] If they had been bitter and not sweet, we would have treated them like toxic drugs, which would have required many robust clinical trials before contaminating our food supply.

So despite the industry denials, all the evidence suggests that ASBs are far from inert and are definitely not a healthy substitute for sugar in drinks or other processed food products. Although, as is often the case, there are likely to be big variations between people, on average ASBs are likely to make you gain weight. Because as they interfere with your metabolism and insulin pathways, they can also increase your risk of diabetes. Worryingly, many ASBs are being used together or combined with other types of sugars known as polyols or sugar alcohols like xylitol, mannitol or isomalt, which are less sweet than regular sugar (sucrose) but have relatively fewer calories. This increasing complexity of chemical mixtures, which our bodies have not encountered before, will confuse our body and gut microbes even more, potentially altering our normal metabolism and behaviours.

Sugar originally comes from plants, so could we dump these lab creations and find a natural substitute? 'Stevia', a new sweetener derived from a South American plant, has been heavily portrayed as a saviour for the 'diet' drink industry. Stevia tastes 300 times sweeter than sugar and was passed as safe in the US in 2008. Coca-Cola created a 'natural' version of their 'chemical' beverage, giving it the name 'Coke Life'. Unfortunately, eventually it had to be discontinued as too many consumers complained of the liquorice smell and bitter aftertaste. This came from stimulating both sweet and bitter receptors at the same time, which many people find an unpleasant sensation. A potential solution would be to combine it with lower-sugar-content drinks to avoid the aftertaste, while keeping calories down. Another would be to use only the sweet non-bitter chemical of the plant. But it is expensive to grow enough stevia to use only this part of the plant and discard the rest. Entrepreneurial companies are now fermenting stevia leaves in giant vats with alcohol and yeast to let the microbes produce the sweet, more rare chemical (Reb M) in large amounts, thereby avoiding the nasty aftertaste.[12]

Fermented or modified stevia could be the holy grail of sweeteners but as usual there is likely to be a catch. The one study I am aware of in humans showed a modest weight gain over twelve weeks, although less than with aspartame.[13] Stevia has some antimicrobial effects, which may be useful in preventing pathogens like listeria and salmonella growing on food but could also damage our friendly gut microbes.[14] It hasn't been shown to cause cancer in rats, but no proper human gut studies have been performed. With anti-sugar pressure growing, versions of stevia will soon be added to nearly every type of processed food containing sugar and it's no surprise that investment is booming in this area. Stevia may be a natural plant – but so is hemlock. It still needs careful testing before we confidently and safely replace the 'poison' sugar without making the same mistakes we did with other sweeteners and additives that fuelled the obesity epidemic in the unsuspecting public.

While occasional exposure to chemical sweeteners in diet drink once a week is unlikely to have much long-term impact, many people who enjoy the taste usually drink two or more cans daily. In fact, some diet-drink addicts regularly have as many as twenty cans a day. Even if you avoid ASBs, you may be consuming them without knowing as these chemicals are commonly added to low-calorie ready meals, cakes, biscuits, fruit yoghurts and desserts. I even found myself drinking a lightly sweetened sports drink after cycling, only to find it had sucralose mixed with the real sugar, with no indication on the front label. Since we are surrounded by processed foods and drinks packed with these potentially harmful chemicals, we should take it much more seriously than we do. This should start with a ban on the use of the words 'diet' or 'low calorie', as these marketing terms ignore the new science which suggests that these artificially sweetened drinks and foods will trick you into gaining weight.[15]

7.
NOT ON THE LABEL

Myth: Food labelling helps us make healthier choices

Food labels supposedly nudge us towards healthier choices, but only a third of Americans frequently pay attention to them, and less than a quarter of Brits bother. Food labelling started appearing on packets in the 1970s when information on the calorie or sodium content appeared on some foods for people with medical conditions who had 'special dietary needs'. Back then, food was generally prepared at home from basic ingredients, meaning there was little demand for nutritional information. Nowadays, around 40 per cent of Americans eat fast food on any given day, with a fifth of meals eaten in the car, while more than half of the food purchased in the UK is classified as ultra-processed. This reliance on convenience foods, combined with our paradoxical increased interest in diet and health, has led to demands for nutritional information. A 2015 global health and well-being survey of 30,000 people found that 88 per cent said they were prepared to pay more for 'healthier' products, including functional foods, GMO-free, and 'all natural' products. However, the food industry have manipulated science and nutritional information to make heavily processed foods appear superficially healthier.

It's clear that food labelling isn't improving our health, as obesity and diabetes levels continue to rise in developed countries. There are few (if any) high-quality studies supporting the use of food labelling, and most of the studies available are either poor-quality or biased due to sponsorship from the food and drink industry. Although I'm strongly in favour of better, more transparent food labelling when it comes to ingredients and the origins of foods, independent reviews suggest that there's now too much information on food labels, confusing and overloading consumers.[1] The food industry latched onto our interest in nutrition, and before long, ambiguous claims such as 'extremely low in saturated fats' were appearing on food labels and in marketing campaigns. Today, pretty much any foods can call themselves 'all natural' or 'superfoods' as these words have no clear definitions or regulation, and are ideal for marketing. A 'natural superfood' like a goji berry, for example, can be sold for ten times more than the humble strawberry despite similar properties.

Kellogg's were one of the first big food companies to legally manipulate the system. They joined forces with the National Cancer Institute in the US and in 1984 used the back of a cereal package to advertise cornflakes as a high-fibre breakfast cereal associated with reduced risk of certain cancers. Their campaign was unimpeded by the FDA, and it paved the way for other companies to follow suit around the world.[2] Nowadays, while there is slightly better regulation, we are still being misled. For example, most cereal bar companies can market what are essentially sugary biscuits as 'high fibre', providing their 20-g bar contains a mere 1.2 g of fibre (recommended daily amount is 30 g). You can call a loaf of bread 'healthy sour-dough' even if it contains less than 1 per cent sourdough flour, while a sugar-laden chocolate bar containing 20 per cent protein can be labelled as being 'high protein'. These pathetically low cut-off guidelines allow the food industry to cash in without adding any real health benefit. Another well-known trick is the 'halo' effect, where a nutrition claim such as something being a 'source of calcium' tricks consumers into thinking a product is healthy, meaning that they overlook the high levels of saturated fat, sugar or salt. They want you to believe that a milkshake labelled as a source of calcium is healthy, despite the fact it's loaded with sugar and you probably don't need the calcium. Many customers scan labels to check for additives and other perceived 'deadly' chemicals such as those with E numbers, but manufacturers have simply given these additives more natural-sounding names such as carrot concentrate or rosemary extract, which are more appealing, despite being equally refined and processed. The E number classification system is simply a way of regulating and identifying over 700 food additives, which have been tested and deemed to be safe by European Food standards, and plenty of everyday foods also have an E number: for example, E160c is paprika and E100 is turmeric. Modified starch is a common ingredient with a friendly name in most processed foods, yet few people know that it has multiple structures and properties to bind different foods together and is made using acids and sugars in highly complex chemical processes. There is very little regulation, and foods that boast about their health on the front of the package are often the unhealthiest. Many labels pretend the food comes from homely-sounding local farms that are purely fictional. Essentially, never trust a food label.

In the US, the Food and Drug Administration rules on food labelling haven't changed for thirty years despite changes in scientific opinion. The agency has suggested a new label to make the calorie content bolder and clearer, despite a lack of any supporting evidence that it helps consumers. Labels also continue to list the amount of cholesterol in the product, even though it is now widely accepted that dietary cholesterol has little, if any health effect.[3] Labels also list the percentage of daily dietary nutrients provided based on a standard and unrealistic 2,000-calorie daily diet, despite the average American consuming nearly double (3,600 calories).[4] There's been even less regulation of food labels in the UK, thanks to skilful lobbying, and mandatory back-of-pack nutrition information only came into force in 2016. The label must include EU-specific information such as the name of food, common food allergens, weight or volume, ingredients, date and storage conditions, preparation instructions, and the manufacturer's details. For obscure reasons, alcoholic drinks are exempt, as are fresh products, including bread if reheated on the premises. Any chemical below 0.1 per cent is also exempt – regardless of how powerful or concentrated it is. Who knows what will happen to labels in our unstable future?

Optional front-of-package food labelling was introduced in Europe in 2013, with about half of countries adopting some system, but with no uniformity.[5] If present, it must include the calorie content per 100 g/ml and in a specified portion of the product, the amounts in grams of fat, saturated fat, total sugars and salt, portion size information, and the percentage reference intake for each nutrient. They are also required to use a colour-coded traffic light labelling system, which supposedly tells you at a glance if the food has high (red), medium (amber) or low (green) amounts of fat, saturated fat, sugars and salt. But the guidelines ambiguously say that we should be picking mainly greens and ambers and fewer reds, but reds don't mean you cannot eat a food. They recommend that you 'keep an eye on how often you choose these foods' but do not define 'how often' nor 'how much'. Despite this, Public Health England proudly claim that their label information supports consumers in making healthier choices. Using this highly flawed system means that Greek yoghurt, cheese, olive oil dressing and nuts would all need to be limited or avoided as they are mainly ambers and reds, despite them been shown individually to be healthy and part of a Mediterranean-style diet – one of the healthiest eating patterns in the world.[6]

Australia uses a voluntary and much simpler front-of-packet health-star rating system. It rates a food's overall healthiness on a scale of 0.5 to 5 stars, based on an algorithm that scores foods overall based on their positive or negative aspects. Food manufacturers and retailers are responsible for its accurate use, which can be manipulated for their products to appear healthier. But even this 'simple' scoring system is confusing consumers, compared to an even more basic colour-coding one like the French nutri-scoring system.[7] Perhaps we should take note from Chile, where one in four school children are classified as obese. In 2016, the Chilean government introduced a starkly simple system of regulating junk food, where a black stop sign was displayed on the front of the packets of all ultra-processed, unhealthy, sugary products. This is a simple and direct way for customers to distinguish healthy from unhealthy food products. Nor can these products be sold or promoted in schools or advertised to children under the age of fourteen. Early signs are that it is positively changing what

mothers are buying for their children. The food industry, however, don't agree and say these in-your-face pack warnings are
perceived as harsh, and reduce consumers' control over food
choices. But in surveys, 88 per cent of consumers approved
and said front-of-pack warnings helped to increase their control over food choices.[8] The simpler the messages, and the less
the food industry controls them, the more effective they are.

Governments support the use of food labelling as a 'soft' way
of educating the public about healthy eating without jeopardis-
ing the money-making opportunities for the food industry, who
argue that they are being 'transparent'. Governments believe
that food labelling works on the outdated simplistic premise
that telling us how many calories are in a doughnut or how
many grams of fat are in a slice of pizza will mean we eat less.
Even fruits contain small amounts of fat and everything is a mixture of macro- and micronutrients, but we have been coerced by
the government and food industry into thinking that everything
is simple and quantifiable. As well as being bad science, it has
been shown repeatedly not to work. If you really want a doughnut, and it is cheap and available, most of us would probably
eat it, regardless of how many calories or grams of fat are in it.

Although meta-analyses of studies show small non-
significant reductions in calories because of labelling,[9] some
studies show labels can have a negative effect, giving the consumer unconditional permission to eat more. A study of over
23,000 US adults found that overweight and obese people not
only drank more 'low-cal' diet drinks than people of a healthy
weight, but they also ate more food.[10] This is probably because
they thought they were being 'healthy' by reducing their liquid
calories, subconsciously giving themselves permission to eat
more. Calorie labelling on menus is mandatory in restaurant

chains and food outlets in the US, and the UK government are planning the same futile initiative as part of their childhood obesity strategy. Leaving a food chain to determine the calorie content of their dishes is like having the mafia run a fair casino. Restaurants systematically underestimate the calorie content of dishes, and manipulate the calorie labelling to make products seem healthier. One large study of 104 US chain restaurants and 250,000 meals saw a modest 4 per cent reduction in calories chosen which disappeared over time.[11] Another study of New York fast-food outlets showed calorie recommendations actually led to a slight *increase* in calorie consumption.[12]

As we've seen, calorie content is a useless way to determine food quality. Almost all junk food relies on sugar, salt and cheap fats, with added chemicals and complex processing to make it tastier and last longer. To distract us from these ingredients food companies market their products as 'low calorie'. A standard portion of nuts provides 147 calories and plenty of fat, whereas a chocolate KitKat contains 106 calories, but the lower calories clearly doesn't make the KitKat a healthier choice. It's highly processed with none of the normal structure of the original products, laden with refined fat, sugar and very little fibre, whereas the nuts are still in their original form, will contain beneficial polyunsaturated fats, an impressive amount of fibre, as well as some plant-based protein. Nuts will also contain fat that is not absorbed and nourish our gut microbes with polyphenols as well as providing micronutrients such as vitamin E and magnesium. Guidelines dumb down the science; not all calories are equal and all fats should not be lumped together, as many are healthy. None of the current labels take into account the quality, nutrient value and diversity of food or its effect on our key digestive organ, the microbiome. Fibre is well known to have health benefits, yet displaying the amounts of fibre in food is only optional in most countries and the healthy plant chemicals (the polyphenols) that are thought to reduce our risk of disease are also ignored. The energy, fat, sugar and salt content are often displayed per 'portion' on packages, but the majority of us eat double the recommended portion size. Most cereal boxes, for example, list a recommended portion size as 30 g, which is a tiny little bowl. Yet most of us probably eat twice the recommended portion sizes used to calculate nutrients.

What's the solution or antidote to bad food labels, then? One approach is a more holistic assessment of a food's nutritional profile, perhaps a front-of-pack warning label telling us how processed or unhealthy a food is, such as the black stop sign system in Chile. Another is to provide information as to where our food was made and how long it's been sitting in storage for, and to forgo useless Use By and Best Before dates, which often result in unnecessary food waste. If we persist with the label, we need to insist on a labelling format that's easy to understand; for example, listing teaspoons of sugar in an entire bottle of soft drink can rather than grams per 100 ml. This will guide us in the right direction. In the meantime, until our politicians act to change the system, your best bet is to judge a food on the quality and variety of ingredients rather than relying on the calorie count or grams of fat from the nutrition label or spurious health claims. Although there are exceptions, broadly speaking, the fewer the ingredients, the less manipulated the product is likely to be. If it contains dozens of chemicals and additives, you probably want to think twice about eating it regularly. If there was better public education about food we wouldn't need to use food labels at all.

8.
FAST-FOOD PHOBIA

Myth: All processed food is bad for us

Globally, fast food generated over $570 billion in 2019, which is bigger than the economic output of most countries. Processed foods get a bad reputation – and, on the whole, this is deserved. Healthy eating guidelines tell us that we need to limit processed foods because they are usually high in calories, fat or sugar, and are nutritionally poor. Too much of these foods, we're warned, will make us obese and give us heart disease and diabetes. The clean-eating movement tells us that all processed food is toxic and suggests that eating half a burger can wipe out a month's benefits of clean plant dieting. When you think of processed foods, hotdogs, burgers, frozen pizza, crisps, Pot Noodles, snacks and sweets might spring to mind, as well as sugary or artificially sweetened drinks. But what about cheese, frozen vegetables and bread? These foods are also processed to some extent, so how do you determine which are bad?

Food processing can be as simple as freezing, canning, baking and drying ingredients. There's clearly a big difference between a microwavable frozen lasagne, a cheap sliced supermarket white loaf and an artisan sourdough bread, so more recently, the term ultra-processed has been used to describe foods that are industrially produced. But not all foods with added ingredients are ultra-processed. The food industry thrives on this confusion and have revelled in the average consumer finding it hard to distinguish between them. The complexity of the food label with calories and macronutrients and health messages adds to the confusion. As an antidote to the food industry, another classification system called NOVA has been developed to differentiate processed foods based largely on the degree of industrial food processing.[1]

The NOVA system splits foods into four categories: the first covers unprocessed or minimally processed foods such as fruit, vegetables, grains, legumes, fish, meats, eggs and milk, which should form the basis of your diet. The second is processed culinary ingredients that make the first category's food taste better. Examples are herbs, spices, balsamic vinegar, garlic and oils. The third category includes processed foods, whereby ingredients such as oil, sugar or salt are added to foods and packaged to enhance or modify the product. Examples are canned fish, smoked meats, cheeses and fresh bread. These foods have been altered, but that doesn't mean they're unhealthy. The fourth and final category describes ultra-processed foods. These industrial concoctions usually contain five to twenty ingredients made not from the whole food itself but mostly from substances extracted from foods or synthesised in laboratories to improve and enhance the product's taste. They have gone through multiple processes such as frying, high-pressure steaming, moulding and milling, and treatment with enzymes and chemicals that don't appear on the label. They contain many added ingredients for cosmetic or sensory reasons and are highly manipulated by taste and feel to be as addictive as possible, so it's difficult to stop eating after just one. The NOVA classification isn't perfect, but most of the criticism has come, unsurprisingly, from the food industry or the many people paid by them.

Certain ultra-processed foods may be marketed as 'healthy', 'natural', 'organic' or 'low-fat', but while these words may describe the original ingredients, they don't refer to the process of how the food was made or its final outcome. Many food companies are reformulating their recipes, adding in 'healthy'-sounding ingredients, such as natural vegetable oil, natural flavourings and wholegrains into the long list of ingredients of snacks to make them seem more nutritious. But the fact is they're still ultra-processed and high in sugar, fat and salt as well as a range of obscure chemicals. The UK consumes the most ultra-processed food out of nineteen countries in Europe, with half of the food purchased being ultra-processed.[2] In contrast, only 10 per cent of food purchased in Portugal is ultra-processed, despite it being a poorer country.

The UK hasn't yet matched the American diet, though. Nearly two-thirds of the food consumed there is ultra-processed, with over $250 billion spent on fast food each year.[3] Many people survive totally on these foods, which are usually full of fast calories and nutritionally poor. Excess consumption correlates with obesity levels and lower socioeconomic status.[4] These ultra-processed foods are increasingly eaten as extra snacks, and these now account for 20–30 per cent of our total energy intake. As snacks become the norm, they disrupt mealtimes and regular patterns of eating. These rapid changes are now global, with countries like China (where snacks were unheard of fifteen years ago) now showing the same levels of consumption of these cheap, highly marketed products. It's an exploding market, with Chinese snack food already worth over $8 billion. It's no coincidence that in the highest-consuming ultra-processed food countries, 75 per cent of the food supply is controlled by just four or five massive processed-food companies and supermarkets.

Not only are these ultra-processed foods affecting our waistlines, there's overwhelming evidence to suggest that they have additional negative effects on our gut microbes as well as our hearts, brains and metabolism.[5] Even if the macronutrients or sugar appear to be at 'healthy' levels in your muesli or granola bar, additional chemicals and enzymes are likely causing health problems. We are consuming large amounts of preservatives, emulsifiers, enzymes and artificial sweeteners, despite a lack of data on their long-term effects, especially in combination. Regulations concerning the use of artificial chemicals are outdated and still focus on whether or not they cause cancer in rodents, while the effects on our gut microbiome are unknown.

As research for my previous book, my son Tom, a twenty-two-year-old student at the time, 'volunteered' for a ten-day fast-food burger and nuggets supersize-me experiment. Just ten days was enough to reduce his microbial richness and he lost 40 per cent of his detectable species. Like many people on low incomes who live on these foods, the lack of diversity of ingredients is obvious: 80 per cent of ultra-processed food is made up of just four ingredients – corn, wheat, soy and meat, with plenty of additives but barely any fibre. His microbe levels and diversity remained depressed for a couple of years. This was possibly because his normal microbes that could feed off the fibre had been wiped out, and were difficult to revive with fruit and vegetables – as he often likes to remind me. A 2014 study of forty-five overweight French men and women confirmed that regardless of body fat, junk-food diets with few vegetables lead to less microbial diversity and more inflammation markers in the blood, which increases the risk of multiple diseases.[6]

The food industry makes huge profits from ultra-processed foods since they use cheap ingredients, and are mostly mass produced and subsidised by taxpayers. The cost of ultra-processed (junk) foods has steadily declined over the last two decades relative to the costs of fruits and vegetables which have increased. Most governments are keen to keep the masses happy by encouraging ever cheaper and poor-quality production of ultra-processed food. They pretend to care about our health by persuading the food industry to slightly reduce the levels of sugar, salt and fat. The food industry loves this approach, as they can take a product and reformulate it with other chemicals and market it now as 'low in salt/fat/sugar'. This outdated belief that food is just the sum of the sugar, fat and salt is complete nonsense. All the evidence shows that regular eating of junk food leads to the greatest increases in weight and ill health compared to other foods.[7]

When the food industry began developing processed foods, they were preoccupied with killing microbes and keeping products on the shelves longer, particularly given the distribution problems in a country the size of America. They knew that fermented products like yoghurt or sauerkraut or pickles (which contain bacteria) kept products fresh, but cakes, biscuits and snacks were more of a problem. They worked out that if you added enough sugar, it would inhibit bacterial growth. Increasing the fat content reduced the water content, which, in turn, reduced bacterial and fungal growth. Finally, on top of the added fat and sugar, the third part of the holy trinity, salt, was added. This also inhibited microbes, preserved the food and extended its shelf life. Together they would produce the conditions for the perfect obesity storm.

There are now a large number of stealth junk foods that are increasingly marketed as health foods. A great example are fruit yoghurts, which have seen massive increases in sales in thirty years, but are packed full of sugar, artificial sweeteners and fake fruit chemicals. To make the yoghurts 'healthy', the manufacturers strip out the fat and replace it with sugar or sweeteners, allowing them to market the yoghurt as 'low fat', despite them being unhealthy. Many popular biscuits, such as Oreos and Digestives, may have been around for over a century but are now made using ultra-processed methods. They contain over ten ingredients, including salt, inverted-sugar syrup and palm oil. I am partial to nibbling the odd biscuit, but many people have them every single time they drink tea or coffee. Other stealth foods are heavily processed in order to be classed as gluten-free or lactose-free, often with misleading 'healthy' labels. Most juice drinks for kids are branded as health foods but are ultra-processed, often with more sugar than colas. For example, Ribena, which proudly markets itself as containing 'real fruit juice', is over 10 per cent sugar, which is more than double the recommended daily sugar allowance for children and equivalent to eleven Oreo cookies.

Before you throw out every single processed food in your home, remember that not all are bad, and several healthy examples include tinned fruit and vegetables, baked beans, cheese and milk. You might not consider canned fruit to be a 'healthy food', as we're constantly told that 'fresh is best'. Most people will crack open a can of pears or mandarins as a last resort, when their cupboards and fruit bowls are empty. Yet most fruit and vegetables are canned close to picking, conserving most of their nutritional value, before being steamed or alkaline cleaned to help remove the skin, peeled and cut, and added to the can. Sugar syrup, natural fruit juice or salted water is added before it is sealed (most of which can be rinsed off prior to serving), steam cooked and cooled to sterilise the contents.

Much of our perception of canned fruit and vegetables comes from our emphasis on the loss of vitamin C through heating. While vitamin C is usually reduced by a third, many polyphenols – the natural defence chemicals found in all plants – are often increased, even after months in a can. The British buy nearly a billion cans of baked beans annually and they are popular in most English-speaking countries, yet it has a reputation as unhealthy. The reality is that this is one of the healthier processed staple foods around, and the beans themselves are highly nutritious. You get around 7 grams of protein and 8 grams of fibre in just half-a-can's serving, which is more than from four pieces of wholemeal bread or six bowls of cornflakes. The amount of sugar was high in many countries for years, but has reduced to around two and a half teaspoons in the UK and Europe, and you can now buy low-sugar versions.

Thanks to the downmarket image of processed foods, most fruits and vegetables are surprisingly affordable when frozen or canned; frozen berries, for example, are two-thirds cheaper than their fresh equivalents. Yet freezing is a great way of preserving the micronutrients found in fruits and vegetables. Most vegetables and some fruits are blanched in hot water for several minutes before freezing to deactivate enzymes that may cause unfavourable changes in colour, smell, flavour and nutritional value. They contain comparable micronutrients to their fresh equivalents, and peas, if frozen quickly, even retain more vitamin C.[8] Most forms of precooked canned pulses or beans are cheap and usually more nutritious than dry varieties, which may have been sitting in a storeroom too long. Meanwhile, people may look down on canned tomatoes as inferior compared to peeled fresh varieties, but there is little nutritional difference, and many canned oily fish are similarly good value and healthy. Canned salmon actually contains more calcium than fresh salmon as the canning process softens the small fish bones, a rich source of calcium, making them edible.

Even nuts, considered to be a virtuous 'superfood' by many, go through several stages of processing before they make it onto the supermarket shelves. Cashew nuts, for example, are steam roasted to loosen the hard outer shell (which contains a corrosive oil harmful to humans). The outer shell is cut off and then the individual nuts are peeled to separate the white nut from the kernel. The peeled nuts are once again placed in an oven to make them crispier before they are packaged and dispatched. Other highly processed but healthy foods include freeze-dried fruits and vegetables that, unlike other drying methods, allows the food to retain its shape and colour and produces a high-quality product. This method is popular for shitake mushrooms and goji berries and involves freezing the product, lowering the pressure and removing the ice. They can be brought back to life for cooking purposes by soaking them for twenty minutes or boiling them.

Virtually all the milk and milk products that we consume are 'processed' in some way. Pasteurisation is a key process in milk-making as it eliminates pathogens and extends the shelf life by applying heat to the milk. Other more highly processed forms exist, such as UHT milk, in cartons that can last for up to a year at room temperature. Fake milks such as soy or almond claim to be healthy but are usually ultra-processed with multiple ingredients. All traditional milk-derived products such as yoghurt and cheeses are processed to some degree. Even your finest artisanal cheese will have been stirred, heated, pressed, and mixed with other ingredients including rennet, salt and other flavourings. But while both are processed there is a world of difference between this quality item and an ultra-processed Kraft cheese slice.

We should resist the urge to be food snobs. Cheap does not always mean unhealthy. Yes, there's overwhelming evidence that eating fresh, unprocessed, whole foods such as fruits, vegetables, wholegrains, pulses and occasional fish and meat will do a lot of good for your body. But we need to keep an open mind and remember that a tin of baked beans or some frozen peas can form part of a healthy and balanced diet. There's a distinct difference between classes of processed and ultra-processed foods and we need to change our definitions and improve our knowledge of food if we are to stand a chance in fighting back the tide of ultra-processing.

If each new ultra-processed food introduced into our diet were a drug made by a pharmaceutical company, and if obesity were labelled a disease, we would have a wealth of data on its benefits and risks. Yet we have no such safeguards. Consider carefully the source and ingredients when buying your food. An apple from a tree versus a bottle of highly processed apple sauce are going to have entirely different nutrient and health profiles. The same goes for a grass-fed steak versus a frozen hamburger patty. If you have trouble figuring out where a food originated because it has been so highly manipulated, or you don't recognise most of its ingredients, it's probably a sign that you should avoid it all together.

We are becoming increasingly reliant on ultra-processed foods purchased outside of the home or delivered. These foods are usually high in rapid calories, fat, salt and sugar and offer little nutritional diversity or benefit. Even making your own versions of your favourite processed foods such as chicken nuggets, pizza and ice cream will be healthier. The one thing that every diet guru, book and plan can agree on is that too many ultra-processed and fast foods shouldn't be eaten on a regular basis. Despite knowing this, we allow these products to be indirectly subsidised and heavily marketed to the poorest, least educated and vulnerable members of our society who become dependent on them. Ultra-processed food companies, seeing saturation in the West, are now focusing on developing countries – with amazing success. In a decade we will look back and wonder why we stood by and allowed a few greedy food corporations to get us addicted to ultra-processed foods, while turning a blind eye to the effects on our health.

9.
BRINGING BACK THE BACON

Myth: All meat is bad for us

For centuries, meat was a rare treat for most people, but now in many countries it is seen as an essential protein staple to build muscle and health. In 1961, regular meat eating was frequent in only a few highly industrialised countries of North America and northern Europe, where rates are only slightly higher today.[1] But rates per person have increased fourfold in the rest of the world and are highly correlated with local wealth and GDP, except in countries like India with its vegetarian traditions.

Alongside this growing consumption, meat has earned itself a bad reputation. We're inundated with messages about deadly red meat causing cancer and its harmful effects on our planet. The World Health Organization recently went as far as classifying red and processed meats as carcinogens, putting them on the same danger level as cigarettes, a move that received much criticism from the meat and agricultural industries.[2] This fear-mongering has led many of us to swap our steaks for roasted cauliflower or beetroot burgers. Plant-based diets and vegan meat alternatives are becoming big business, and the meat-substitute industry is growing at around 15 per cent a year, though it is still only tiny compared to the trillion-dollar real meat business. This could change, however. The small US company Beyond Meat, with funding from Bill Gates and Tyson Foods, was already worth $8 billion at the end of 2019, and their rival Impossible Burgers, with funding from Google Ventures and other Silicon Valley firms, is close behind. Giant food companies like Nestlé are starting to make their own meat-alternative brands.[3] In 2018, one in five Britons were already reducing their meat intake, and we are being encouraged to have several meat-free days per week, with global initiatives such as 'Meat-Free Monday' or 'Veganuary' being headed up by big-name celebrities. This has had a knock-on effect for the meat industry, and the UK saw sales of beef, pork and lamb dip by 4 per cent in 2016, while US meat consumption has plummeted by 15 per cent in a decade. Even in China, where 98 per cent are meat eaters, 36 per cent of city-dwellers surveyed in 2017 said they were trying to eat less pork. Within Europe, the UK has one of the highest levels of non-meat eaters, with about one in six people self-reporting as non-carnivores, and about one in fifty being vegan. This is four times as high as France and about eight times higher than the meat-loving US. People choose not to eat meat for many reasons, but the most common are concerns about animal welfare, the environment and, increasingly, for the health benefits that are supposed to come with avoiding meat.

But real meat isn't as bad for you as we've been led to believe. Meat is made up of water, protein and fat, plus a few carbohydrates, iron, zinc and B vitamins. When we talk about protein in meat, we are referring to the muscle the animal needs to move. Red meat contains high amounts of the iron-rich myoglobin protein, which gives it its characteristic colour, and other nutrients such as selenium, zinc and B vitamins. White meats such as chicken and turkey contain less myoglobin, iron and zinc, and the meat is more tender and lower in fat. Humans are mainly made up of red muscle, which helps us to run marathons; pigs, in contrast, don't run far and are mainly white muscle; chickens are a mixture of both, with their dark legs in constant use, while their white breasts for wings are hardly used at all.

Much of the confusion about the dangers of meat comes from our fixation on dietary fat and heart disease. As we've seen in chapter 4, the idea that fat is deadly arose in the 1960s, when cholesterol in foods first got the blame for heart disease, although this was a theory later disproved. Next up, saturated fat got a battering from observational studies, and supported by short-term clinical trials mainly conducted in Western populations, which found that LDL cholesterol – a marker of heart disease – is associated with increased dietary saturated-fat intake. Dairy and meat are indeed a source of saturated fat (as are nuts, olive oil, coconut oil and other so-called 'health foods'), and we have developed a prejudice against them as a result. Yet a recent observational study of 135,000 people in poorer countries (PURE) found that eating relatively higher amounts of saturated fats from meat and dairy was associated with lower mortality compared with those eating more carbohydrates.[4] So meat eating has to be taken in context and, in this study, the amounts of meat and dairy eaten were much lower than typically seen in the West. Meat eating is also associated with being wealthier, which, although adjusted for, could still have biased the mortality results. Nevertheless, this study clearly shows that our previous assumptions should be challenged. UK and US dietary guidelines still encourage us to reduce our saturated-fat intake, aiming for less than 20 per cent of total energy per day, replacing red meat with leaner cuts of meat such as chicken and turkey, trimming fat off our meat and opting for reduced-fat versions of sausages and mince. Unfortunately, many of the studies looking at meat consumption and saturated fat have not considered the complexity of meat or the fats in, for example, the pastry in the meat pie. Yet the belief that overconsumption of red meat causes heart disease remains a core tenet of nutritional advice.

Look more closely at the data behind this nutritional advice and, as so often, things start to get more complicated. Observational studies of over a million people from the US, Europe and Asia have confirmed that regularly eating red meat increases mortality and heart disease by a small amount. The increased risk per extra meat portion per day was around 10–15 per cent, increasing to 30 per cent with processed meats.[5] There was a modest increased risk of cancer of around 15 per cent. The risk estimates are more consistent in the US, possibly because Americans eat a lot more meat (Americans annually eat 127 kg vs 84 kg in the UK). Based on this data, it was estimated that halving European meat intakes or reducing US levels by a third would reduce premature deaths by around 8 per cent. But another meta-analysis hit the headlines in 2019 looking at exactly the same data as four previous groups. Its conclusion was very different. The Canadian authors said there was no good data suggesting eating red or processed meat was bad for you.[6] I was surprised by the results, and the controversy they provoked,[7] but when I was asked to write about it with my American colleague Christopher Gardner for the *BMJ*, we found out the authors had selectively removed most of the epidemiology and short-term studies, leaving only a handful that showed no effects. It also emerged that the senior author had received funding from ILSI, a front organisation for the food, drink and meat industry and a Texan university, and had previously published similar conclusions about the safety of sugar.[8]

This is not to say the data in favour of reducing meat consumption are clear-cut either. When you look at Europeans and Asian populations separately, it's hard to see any effect of red meat on deaths from heart disease, perhaps because of other healthier plant foods they eat with meat. Studies of 300,000 Asians from Japan, China and Korea, where the amount of meat eaten is much lower than in the West, suggest that eating red meat actually *reduces* the risk of mortality from heart disease in men and of mortality from cancer in women.[9]

More powerful randomised clinical trial data of the effects of diet changes have been overlooked; for example, the eight-year colon cancer prevention study of 1,000 people, or the 38,000 women in an eight-year cancer prevention trial of low-fat diets.[10] These participants, as part of other diet changes, reduced their meat intake by around 20 per cent. Both studies found no decrease in cancers or mortality in those giving up red meat.

In 2011, following a report that linked red and processed meat to colorectal cancer, the UK government recommended that people should reduce their intake to 70 g. In a further blow to the meat industry, in 2015, the World Health Organization Working Group of twenty-two scientists assessed over 800 studies and produced a report about red and processed meats that upset a lot of people around the world.[11] It said that all the epidemiological and lab toxicology evidence pointed to red meat being a 'possible' carcinogen and processed meat being a 'definite' carcinogen, with an 18 per cent increase in colon cancer in people eating an equivalent of two rashers of bacon a day. There were several problems with this report common to most areas of nutritional research. First, it was based on weak observational epidemiology data that cannot separate cause and effect. Secondly, it omitted to discuss more powerful (and negative) clinical trial data from people giving up meat in the cancer prevention trials. Other recent larger meta-analyses of observational studies of over a million people have come to the opposite conclusions, and have shown no effect from red meat on cancer and only small effects from processed meat.[12] Thirdly, the research committee were not completely independent and unbiased, being composed of a majority of vegetarians with a track record of red meat research, and finally, the report was never released in full as promised, or peer reviewed, thus preventing normal scientific criticism.

Putting red meat in the dock alongside tobacco and plutonium was a ridiculous scare tactic. Clearly, eating the occasional burger does not pose the same risk as smoking a packet of cigarettes – dose is important. The research committee – and most journalists reporting their findings – forgot, as they usually do, to provide context. They should have pointed out that the cancer risk of eating a hundred rashers of bacon every day was equivalent to smoking every day, or that the average cancer risk of an Italian meat eater equated to only smoking three cigarettes a year. The WHO report also didn't differentiate between quality and type of meat. For example, eating stacks of quarter-pounders high in salt, saturated fat and additives isn't the same as eating a small, organic grass-fed steak, which has much higher levels of potentially heart-healthy omega-3 fatty acids.[13] The 'bacon causes cancer' headlines should therefore, be taken with a large pinch of salt.

If we should be sceptical about cancer risks associated with meat eating, what about environmental considerations? There is an increasingly accepted school of thought that a predominantly plant-based diet can save the planet and feed the world. The EAT-Lancet report published in early 2019 by an international group of academics claims that people must drastically reduce their meat and dairy consumption not only to be healthy but also to reduce greenhouse gas emissions.[14] The report published, for the first time ever, a set of specific, numerical targets for national dietary guidance that supposedly support optimal nutrition and planetary boundaries. Their prescriptive 'Planetary Health Diet' recommends eating daily 13 g of eggs (the average egg weighs 50 g), 14 g of beef, lamb or pork (a small steak weighs 85–100 g), along with other rigid recommendations for grains, fruit, vegetables and dairy. This is diet designed by computer algorithm – it is tricky to eat a third of an egg or a tenth of a steak. Once again, the focus was on the quantity of meat eaten, not its quality. This was one of the first instances of a group of scientists funded by charities trying to compete with the food industry on the same terms to influence consumers. They even used a PR firm to promote their findings.

The report was influenced by three recent studies. The first was a 2018 study surveying 40,000 farms that reminded us that one of the biggest modifiable factors in greenhouse gases and global warming is land for agriculture, which accounts for 25–30 per cent of the problem. Livestock for meat and dairy accounts for 83 per cent of all land use, equivalent to the combined area of the US, Europe, China and Australasia. Around

95 per cent of all mammals on earth are domesticated in farms for human consumption in just a handful of species. Of all the domestic animals, cows bred for beef are the most inefficient in terms of protein production and emissions, being on average (globally) about seven times less environmentally efficient than pork, and around ten times less efficient than chickens, and about thirty times less efficient than the equivalent protein from nuts or tofu.[15] Although there are huge differences in environmental impact of the farms within each category, which can vary about fivefold, even the most sustainable beef herds are four times less efficient in land use than the least efficient nuts or pulses. Nevertheless, by simply stopping eating below average sustainable beef from the worst producers we could reduce land use by three-quarters. Some of the worst offenders are global companies who sponsor cutting down areas of Brazilian rainforest the size of Manhattan every day, just to provide more grazing land or soy or corn feed for cheap beef.

A second report had concluded that to reduce climate change through food and halve livestock emissions, the average human would need to eat 75 per cent less beef (90 per cent less in the US) and half the number of eggs, while tripling consumption of beans and pulses and quadrupling consumption of nuts and seeds.[16] The third study was a modelling study that looked at the effects of taxing red and processed meat on health and economic outcomes.[17] It suggested that diseases related to meat consumption kill 2.4 million people a year and cost $285 billion in health-related costs, and that taxing red meat would save 222,000 lives per year. The authors consequently proposed a US tax of 163 per cent on processed meats and 34 per cent on red meat, while in the UK, there have been calls for a 79 per cent tax on processed meats and 14 per cent on red meats. There was a strong reaction to this; the then UK environment secretary Michael Gove described it as 'the worst example of a nanny state', but we no longer hear such objections to alcohol or cigarette taxes. Other countries are currently debating similar taxes, although Germany recently withdrew plans for a *wurst* tax due to a popular backlash. A novel way to reduce harm to the environment without having to give up your burgers is to eat stem cell lab meats – a rapidly expanding industry which so far has created relatively similar-tasting products using vastly less water and land, and generating less greenhouse gases. Insect meats also boast a similar reduction in harm to the environment. Both will likely be on our plates within the next five years.

Agricultural organisations such as the US Animal Agricultural Alliance and UC Davis Department of Animal Science have heavily criticised the EAT-Lancet report, saying that it ignores the nutritional benefits of meat, greatly exaggerates negative health outcomes and ignores regional meat industry efficiencies. The beef industry, especially in the US, has improved its efficiency to produce protein by over a third in the last few decades and reduce its environmental impact, although this so-called 'efficiency' comes with added welfare and hidden environmental concerns.[18]

The WHO and EAT-Lancet reports assume no negative health outcomes from greatly reducing meat consumption. In Western diets, red meat provides a large amount of our daily protein, vitamins, and minerals like iron and zinc. Cutting out such a large food group as red meat without careful planning and appropriate substitutions can put you at risk of nutritional deficiencies.

Almost half of teenage girls in the UK have iron levels below the lowest recommended nutrient intake, and 5 per cent are estimated to have iron deficiency anaemia.[19] We're also seeing increasing rates of iron deficiency in toddlers and young children who are raised on plant-based diets. These rates are likely to increase if more people exclude red meat, which is the richest dietary source of iron, from their diets. You can meet iron and other nutrient requirements on a plant-based diet, particularly in the US where many foods are fortified, but you absorb less iron from plant foods as opposed to meat, and the fortified foods often contain elemental iron (i.e., iron filings) which is less well absorbed, so you need extra advice or nutritional knowledge to ensure you're getting enough. Many vegans and vegetarians run into other nutritional problems as meat contains vitamin B12, zinc and selenium, which are hard to find in plant sources, and many vegans resort to nutritional supplements. I had a brief spell of veganism, which lasted for about six weeks until I went for a medical check-up and noticed my blood levels for vitamin B12 (which is essential for brain health and only found naturally in animal products) were low. I resorted to supplements and even vitamin injections until I had an epiphany and realised that taking artificial supplements was going against the healthy lifestyle I was striving for. I now eat high-quality red meat a couple of times a month and my vitamin B12 levels are back to normal.

Small amounts of higher-quality meats may even be beneficial for our health.[20] Some emerging research is suggesting that red meat consumed in small amounts reduces our risk of mental health problems, such as depression and anxiety disorders. One study of 1,046 Australian women found that lowering their red meat consumption nearly doubled the risk for major depressive and anxiety disorders, while red meat intake below the recommended guidelines for Australians (65 g of red meat per day) was associated with increased psychological symptoms and a higher likelihood of an anxiety diagnosis.[21] This needs repeating but the authors concluded that red meat consumption may play a role in mental health, independently of overall diet quality. The type of meat that you choose may also affect your mental health: grain-fed beef, common in the US, isn't nearly as nutritious as grass-fed beef, which is much higher in the omega-3 fatty acids that have been linked to improved mental health.[22]

How we cook our meat also appears to be important. We now know that the way that food is cooked alters the structure and thus the extractable energy – so, for example, the longer red meat is cooked for, the more calories it provides, although you lose some beneficial antioxidants. Acrylamide, a chemical which forms as a result of burning foods such as sausages, often hits the headlines. It is produced when the amino acid asparagine combines with some natural carbohydrates. This happens when overcooking anything from toast to sausages or steak. The UK Food Standards Agency warned its vulnerable citizens against eating burnt sausages via a major media campaign in 2017. This health warning came about because acrylamide was categorised as 'carcinogenic' by a WHO/IARC (International Agency for Research on Cancer) committee. This scare story was in fact based on a few lab animal experiments using massive doses of the chemical, not remotely equivalent to a burnt sausage, and an observation that cattle grazing near a tunnel in Switzerland developed a mysterious illness that was traced to large amounts of the chemical in a local river. Despite the scare, a review of more relevant human studies showed no clear effect on cancer.[23]

The same goes for polycyclic hydrocarbons, which are produced when cooking over a direct flame or barbecue. Suggestions that they cause cancer comes from lab studies plus recent observational data of high rates of cancer in fire-fighters working with smoke.[24] These findings are unreliable and based on small numbers. Unless you routinely abuse meat by incinerating it on the barbecue, you shouldn't worry. We are all exposed to hundreds of nasty chemicals every day and it is only when they are combined with others in high doses that we get significant health problems. I am not advocating eating burnt meat every day, of course, not least because over-cooking destroys most flavours of the meat; but it is not something to lose sleep over.

Headlines about the dangers of meat consumption tend to focus on red meat, but what about other meats? There is in fact little evidence that red is significantly more harm-ful than white or even (as we will see later) fish meat. There is also no uniform definition of red meat, with pork usually being included in nutritional definitions, but not in gastro-nomic ones. Many people are switching from beef and lamb to cheaper and often leaner whiter meats like chicken, turkey and pork, but this isn't necessarily a healthier solution. Pork sales have increased since 2011 in countries like the US, buck-ing overall trends despite no evidence it is healthier than beef. Most, but not all, observational studies have shown slightly reduced mortality risks of 5–7 per cent if you eat lots of white meat like chicken or fish, although eating processed versions (such as chicken nuggets and scampi) increases the risk. The difference between white and red meat data are puzzling and probably exaggerated. Some of the difference may just be due to inaccuracies in how the data is collected and what else people are eating.[25] A 2019 clinical trial comparing 113 Americans eating chicken or beef for four weeks showed no difference in heart-risk markers once fat levels were adjusted for.[26]

There are other costs to eating whiter meats too. Industrial farming of pigs with weak immune systems kept in cramped conditions has led to epidemics of African swine fever across Asia, causing shortages and higher prices with an estimated 350 million dead pigs in China alone by 2020. Ultra-cheap battery chicken has become readily available, at the expense of nutritional value and quality. If you're eating cheap supermarket chicken, chances are it was battery farmed, raised in cramped conditions where infections spread like the plague, and widespread use of chemicals and pesticides is routine. In the UK, with the uncertainties of Brexit and trade wars affecting our food supply, there has been much alarm about the import of chlorinated chicken from the US. Chlorination of chicken removes harmful bacteria, but it's been banned in the EU since 1997 over food safety concerns, a move that stopped virtually all US chicken imports. The European Food Safety Authority have agreed that the evidence suggests the chlorine wash itself is not harmful.[27] But the concern is that treating meat with chlorine is a symptom of everything that we are doing wrong in the poultry industry; most farmers don't care about infection rates in their chickens and resort to chlorine as a temporary solution. Prior to this, mass use of antibiotics kept infections at bay and promoted growth of livestock, but this led to record levels of antibiotic resistance in poultry, which perpetuated the global catastrophe of antibiotic resistance in humans – one of the biggest threats to human health – and led to a 2018 ban on pre-emptive antibiotic usage in livestock in the EU.[28] Antibiotics continue to be widely used in poultry production in many countries such as New Zealand, India and China. In the US, rates are reducing, though in 2018 around 50 per cent were still being given them routinely. But before anyone in the UK feels too smug, battery farms still routinely use non-banned antibiotics called ionophores to fight parasite infections in the chickens, and most British chickens and their packaging are caked in salmonella and campylobacter, which cause around 300,000 bouts of food poisoning per year.[29]

When it comes to red and processed meats, all of us should seriously consider being flexitarians – if not for the sake of our health then to reduce global warming. Reducing your meat consumption, especially poor-quality, low-sustainable versions such as grain-fed beef, could be the most important thing you can do for the planet. Around 2 billion people on the planet exist without eating any meat, so clearly if you are not used to having it, meat is not essential. Since the data shows that mortality is slightly increased in heavy red-meat eaters, and more definitely increased in processed-meat eaters, it is sensible to assume we are eating too much for our own good, either because of something in meat itself, the consequent lack of vegetables we could be eating instead or other dietary habits in meat eaters. Yet, simply switching to cheaper white, lean meats such as mass-produced, battery-farmed chicken isn't the answer. Eating too much of anything won't be good for you and we have become so seduced by the supermarkets, with their ultra-low price and mass-produced meat, that we now eat meat daily, with little if any preparation or thought. We have forgotten how to eat a diversity of animals, many of which are discarded or used as dog food, including rabbits, hare, game, birds, ducks, goats, and many unfortunate male animals that are too costly to keep alive. By turning our back on different cuts of flesh and organs filled with nutrients, that our grand-parents ate, we are wasting our planet's limited resources.

The price of virtually all meat has reduced dramatically around the world, as it takes less time and money to produce it now than ever before. But unless we rapidly change our habits, we will run out of land to grow all the crops that are needed to feed all the animals we need to kill regularly for meat. Meat is heavily subsidised in most countries, especially in large industrial-scale mechanised units, meaning it's affordable to most, while fruit and vegetables and rich sources of protein like legumes are not. Industrial meat production also has hidden additional environmental and pollution costs which can double its real costs. I believe we should pay the proper price for meat, and the easiest way to do this is via extra taxes which would enable us to subsidise cheaper fruit and vegetable production. But until our governments are bold or brave enough to introduce these measures, we can all make a difference by treating meat as a luxury, bulking out meat dishes with beans, pulses, vegetables or mushrooms and having a few meat-free days each week. We should all pay as much as we can for higher-quality cuts of more sustainable grass-fed grazing animals, which have beneficial effects on the soil and for organic farming. And we must remember that context matters: eating a homemade hamburger made from high-quality minced meat from a known source is always going to be preferable – for your body and for the planet – to eating an anonymous patty. Above all, while occasional small amounts of top-quality meat are probably good for you, as we'll see in chapter 11, eating a vegan or vegetarian diet isn't necessarily healthier.

10.
FISHY BUSINESS

Myth: Fish is always a healthy option

If meat isn't necessarily bad for us, what about the equally widespread notion that fish, another form of animal flesh, is good for us? In the 1930s, eating fish helped to eliminate childhood rickets, a major public health epidemic causing stunted growth, leading to long-lasting hype over its health benefits. Children would line up at school to receive their daily dose of goodness: cod liver oil, malt and a side of stale bread to take the taste away. This daily ritual, along with fortification of certain foods including dairy products with vitamin D, eradicated this significant health problem within a decade. Fish has since been given a 'superfood' status despite it being another meat, but, contrary to modern belief, this had nothing to do with coronary heart disease. We're constantly inundated with messages telling us that fish is one of the healthiest foods on the planet, a low-calorie and high-protein food, with oily types rich in omega-3 fatty acids (also known as DHA or EPA or fish oil), which is supposedly good for our hearts and brains. Our obsession with fish is something that once again has been manipulated by the food and supplement companies, and the fish oil supplement industry is worth an eye-watering $30 billion. Almost 10 per cent of Americans and 20 per cent of Britons take a daily fish oil supplement (it's the commonest dietary supplement), and the UK spends £2.8 billion per year on fish. But there's bad news for fish fans: it isn't as good as we've been led to believe.

For decades we have been preaching that eating fish is crucial for good brain development, children's academic performance and reduced risk of disease, while there's lots of observational data showing it's good for your cognition and memory.[1] I used to coax my son into swallowing fish oil capsules for years, only to find he was hiding them behind the kitchen cupboard. One large community study followed people aged sixty-five and over for six years and found that fish consumption may potentially be associated with slower cognitive decline with age.[2] Unfortunately, observational studies can, at best, suggest an association between two variables (here, fish consumption and cognitive decline), but the observed effect – in this case, slower cognitive decline – could be due to other factors, such as diet and lifestyle, namely fruit and vegetable consumption or daily exercise.

In the early 2000s there was a huge publicity campaign encouraging parents to give their children omega-3 supplements. Omega-3 fatty acids are found in fatty fish such as salmon and mackerel, and in walnuts, flaxseed, algae and fortified foods. There are two main types of omega-3 fatty acids: long-chain omega-3 fatty acids known as EPA (eicosapentaenoic acid) and DHA (docosapentaenoic acid), which are found in fish and shellfish, and short-chain omega-3 fatty acids known as ALA (alpha linoleic acid), which are

found in plant foods such as flaxseeds, chia seeds and walnuts. Although ALA is important, DHA and EPA have more
potent health benefits. The main issue with ALA is that
it must be converted into EPA and DHA to have the good
effects attributed to omega-3s, but the natural process is slow
and inefficient. Only a small amount (about 10–15 per cent)
of ALA gets converted, which is probably why studies have
shown that dietary intake and blood levels of EPA and DHA
are much lower in vegans than omnivores.[3] We don't know
the clinical significance of this as vegans don't tend to dis-
play signs of deficiency, however vegans need to eat plenty
of ALA-rich foods to get the same benefits of omega-3s from
fish. Some vegan organisations encourage vegans to take
vegan microalgae supplements daily, as they provide both
EPA and DHA.

The brain needs the main omega-3 fat DHA to develop normally, so it made sense to think that children with low dietary
levels would benefit from a top-up from fish oil. It turns out
we were misled, and meta-analyses of many randomised clinical trials have shown no consistent effect from supplements
on children.[4] The Norwegians are keen to push the benefits of
oily fish rather than supplements, as they produce so much of
it, but even in their own trial feeding 214 preschool kids with
mackerel or herring instead of meat for lunch for four months
failed to find any improvements in cognition.[5] Many expectant mothers take omega-3 oily fish supplements during pregnancy thinking that these fats will help to boost their baby's
brain development. But a recent clinical trial of 259 pregnant
women followed up their babies until they were seven years
old and found that fish oil supplementation during pregnancy
did not lead to brighter children.[6] The evidence for omega-3

supplements being beneficial is non-existent, and the evidence for eating whole fish not much better.

The health benefits of eating real fish is hard to study and not conclusive; an observational study of 500,000 Europeans followed for fifteen years found no overall benefit on mortality, and excessive fish consumption may actually increase mortality slightly.[7] A recent summary of all the twenty-nine studies performed to date showed a very modest 7 per cent reduction in mortality per weekly portion compared to a 24 per cent reduction for nuts.[8] So even if we followed the current dietary recommendations to eat two to three portions of fish per week, it probably won't help us live that much longer. Besides, vegans, despite avoiding fish, live longer and have fewer health problems than omnivores.[9] That said, many fish-eating populations are in good health, particularly those who consume a Mediterranean or Asian diet. There may be individual differences that mean eating fish provides benefits in some people, and this could be dependent on their unique gut microbes. However, many of the remarkably healthy villages packed with centenarians in the mountains of Greece and Sardinia eat little fish.

When people talk about the magical properties of fish, they are usually referring just to the omega-3 fatty oils, which are most commonly found in salmon, trout, pilchards, herrings, sardines, sprats and mackerel. For years, this type of fat has been cherished, along with the idea that eating concentrated amounts is healthy. Omega-3 supplements have been widely recommended for people who find it hard to eat the government-advised two to three portions of fish per week. But do they actually work? In 2002, the influential American Heart Association recommended them for the prevention and

treatment of most heart conditions, and they have been prescribed around the world in huge numbers. But fifteen years
later, the same committee reviewed evidence from a review
of twenty new randomised clinical trials of supplementation.
These larger, more recent studies revealed zero benefits to taking supplements, with little to no evidence that they prevent
heart problems – the only exception being that they may be
worth taking for the six months after surviving a heart attack.[10]
Results from a 2018 US review of ten large, high-quality, year-
long studies were even clearer. There was no effect whatsoever of fish oil supplements on risk of heart disease or stroke,
and they should not be recommended.[11]

A recent UK review combined seventy-nine randomised
trials involving 112,000 people, and concluded the same: taking
long-chain omega-3 (fish oil, EPA or DHA) supplements does
not benefit heart health or reduce our risk of stroke or death
from any cause.[12] This was reinforced in 2019 by a massive US
trial of 25,000 people for five years which was also negative.[13]
The evidence is so compelling that UK clinical guidelines for
prevention of cardiovascular disease (NICE) no longer recommend omega-3 fatty acid compounds. They have also dropped
the statement that eating fish helps to prevent another heart
attack. The supplements are also marketed for preventing
dementia and helping arthritis. But other large, independent
reviews have found no significant benefits for Alzheimer's,
memory loss and osteoarthritis.[14] At one point, supplements
were prescribed directly by doctors, but now this is stopping,
which has had a knock-on effect on the massive fish oil supplement industry, who have experienced a recent decline in
sales. If vegans can survive without fish (and there are plenty
of plant sources of omega-3s), why can't the rest of us?

Western governments drum into us that we must cut down on our meat intake and instead eat two to three portions of fish per week if we wish to reduce risk of disease. This is despite a complete lack of evidence of benefit. While there's also no evidence so far that fish does our health any harm, there is a clear downside to eating more of it. Our oceans are struggling to meet the demand, not helped by increasing trends for cheaper farmed seafood. There is already pressure on stressed fisheries, which are struggling to meet our extreme and unsustainable demands for fish. Globally we already eat on average 20 kg per person of fish per year; many species are collapsing, reducing biodiversity along with increases in nutrient deserts in our oceans. If the entire population were to follow the government guidelines, the oceans wouldn't be able to sustain it, particularly as the population continues to grow. We're already seeing an emerging trend whereby salmon is the new chicken. It's become much more affordable – at a real cost.

Farmed fish used to be a rarity but now accounts for the majority of fish we eat globally. Just because a fish is labelled as 'Scottish salmon' doesn't mean it's been handpicked from the scenic lochs and rivers of Scotland. Cheaper supermarket fish will usually have been farmed, meaning it won't bring the same benefits as line-caught fish would. Most of our salmon, trout, carp, tilapia, catfish, sea bass, sea bream, hake, prawns and shrimps are now farmed this way, and often travel thousands of miles to get to the fish counter. The more we farm fish, the more we endanger the wild ones, not just from rogue ones escaping and terrorising the locals, but from their carnivorous habits. Much of the fish feed used for farmed fish comes from killing other smaller fish, like anchovies and sardines, to provide protein and extra omega-3. As well as the ground-up smaller fish, there is fish oil, soy, GM yeast, chicken fat and sometimes ground-up feathers in the fish feed. To make dull farmed salmon appear more like the healthy wild salmon that feed off shrimps, algae and krill, they are fed a pigment (astaxanthin) to change their dull grey flesh a pink colour. Aquaculture is under pressure to be more sustainable and in 2015 you had to kill 1.3 kg of wild fish to feed 1 kg of farmed fish like salmon.[15] Farming crustaceans like prawns has a greater effect on greenhouse gases than the equivalent amount of pork, and even other farmed fish leave a bigger footprint than cheese, eggs or chicken.[16] If trends are to continue, we will cripple our ocean ecosystems as well as overheating the planet.

As well as using up our natural resources, some intensively farmed fish should come with a health warning. First of all, there is the routine use of high levels of antibiotics to help the fish grow and avoid infections. As the fish are housed close together, infections are common and spread rapidly, causing many less regulated countries to use tonnes of antibiotics. Chile is one of the world's biggest fish exporters, and alone used 300,000 kg of antibiotics in 2014, which not only led to many fish becoming resistant, but by entering the human food chain, it further perpetuates human antimicrobial resistance, one of the biggest threats to global health. Fish farmers argue that the antibiotics used in farms wash out of the fish flesh before they are eaten. But a 2014 US study looked at antibiotic levels in twenty-seven fish samples (including the commonest fish: shrimp, tilapia, salmon, trout and catfish) from eleven countries, which were bought in shops in California and Arizona. Antibiotics were detected in three-quarters of samples, including those labelled as antibiotic-free. Most reputable fish farms have stopped using antibiotics routinely, but we live in a global economy and a 2017 study has shown that some industrial mass-produced fishmeal products from around the world contained considerable levels of antibiotics and, more worryingly, hundreds of antibiotic resistance genes. Investigations showed that these antibiotics and genes can transfer from the feed to the fish and then into humans.[17]

Antibiotic use has probably risen due to an unlikely little critter that loves closely packed animals: the sea louse. These flesh-eating sea lice attach themselves to salmon, injuring or killing them off in the process. While we're told reassuringly by fish farmers that they're harmless to humans, they are currently killing one in five salmon, which costs the industry over £1 billion a year. The worldwide supply of salmon fell almost 10 per cent last year, with Norway, the largest producer in the world, especially hard hit. Sea lice have already colonised an estimated 50 per cent of the 250 Scottish farms and many others worldwide. If the infected fish escape their pens with lice attached, they can introduce the infection into wild fish, exacerbating the issue.[18] Control is difficult and use of pesticides and antibiotics have been ineffective. Canadian epidemics led to farms using high levels of pesticides, which the lice then became resistant to. A popular natural solution in 2015 was to add special fish to the tanks called wrasse or 'cleaner' fish shipped from English waters that eat the lice off the salmon. This appears to work only when numbers of lice are small, but the large-scale problem has meant that companies are now using hundreds of thousands of tonnes of hydrogen peroxide chemicals to control them. Moving the fish to colder deeper water with more space is not yet economical, but the Norwegians are now investing in giant offshore structures that look just like oil rigs to house the fish.

Legally, Scottish fish can be sold if it contains up to eight lice per fish, but the reality is that Scottish salmon sold in supermarkets often contains up to twenty times the legal amount.[19] The legal limit in Canada is three lice per salmon, although some farms produce salmon with as many as thirty lice per fish. The Scottish farming sector are urging politicians to lower the acceptable level of sea lice on Scottish farmed salmon to tackle this growing epidemic. This is a worldwide problem. Fish farmers consider sea lice to be the biggest threat to their industry and it is driving up costs for consumers, yet certain companies are regularly breaching the acceptable levels of sea lice, or pesticides, all in the name of profit. Countries such as Canada, Scotland, Norway and Chile have been particularly hard hit, and Canadian fish farms have had to resort to using expensive hydrogen peroxide to tackle a recent outbreak, which affected 50 per cent of fish at one of the country's largest salmon farms in 2018. Instead of buying poor-quality farmed salmon, the Environmental Working Group suggests eating other affordable and sustainably produced species, which are rich in omega-3s, including anchovies, sardines, farmed trout and mussels.

Choosing your ideal fish and seafood isn't as simple as it sounds. Fish is one of the easiest meats to fool consumers with, and mislabelling is a major global problem as names of fishes vary in different regions and countries. Some fish names are entirely fabricated to sound better to consumers and sell more, including Pacific rockfish, which is an unclassified, previously discarded fish; or the ugly-looking Patagonian toothfish, which used to be rejected until it was rebranded as classy Chilean sea bass in the 1990s and became a runaway success. Mud crab and monkfish have also been given makeovers and rebranded successfully in the USA. Rebranding is one thing, and can be beneficial if it serves to make discarded fish edible, but deliberate fraud is different.

Fake fish is a big business, and you probably aren't getting what you pay for even in the fanciest of restaurants, particularly in the USA where fish fraud is a huge problem. A new study by the ocean conservation group Oceana found that nearly a quarter of the 400 fish sampled from 277 locations in the District of Columbia was not what it said on the label or menu.[20] Favourites like sea bass and snapper were actually cheaper poorer alternatives like farmed tilapia. A global report of fifty-five countries and 25,000 samples suggested the problem affects one in five fish sold, with over half of the substitutes being cheaper, and potentially harmful, farmed fish like Asian catfish, which are often fed growth hormones.[21] In the UK, expensive cod for fish and chips is often replaced by cheaper pollock, and in a survey using DNA testing between 2013 and 2015, half of raw fish in sushi in the Los Angeles area was mislabelled, with snapper and halibut replaced with cheaper flounder, often without the restaurateur knowing.[22] Tuna is a particular problem because of the demand and high prices at the top-end. US surveys reported over 70 per cent of sushi tuna was fake, and the use of 'white tuna' is common in restaurants. The problem is that white tuna doesn't exist. It's actually escolar, a cheap fish nicknamed 'ex-lax', which is banned in Japan and Italy due to unpleasant effects on the gut.[23] Unless you know your fish or sushi restaurant well, it may be best to avoid the tuna.

We're often led to believe that when it comes to animal products, fresh is best. But buying frozen fish may not be a bad idea if you're keen to avoid other parasites that commonly affect fish, such as tapeworms and Anisakis larvae.[24] The good news is that infections in humans from these parasites are rare and can be treated by a short course of anti-parasitic medication. However, you can avoid this risk altogether by freezing a fish for twenty-four to seventy-two hours, as this should kill any parasites. The Japanese worry that freezing sushi may alter the taste, but a randomised control trial showed this not to be the case, which means that freezing may offer a solution to reducing risk of certain parasitic infections, such as anisakidosis, which is often found in Japanese sushi.[25] Frozen at the point of harvesting, frozen fish is also likely to be fresher, not to mention cheaper, and even the 'fresh' fish on most supermarket counters have often been previously frozen. You'll pay up to 40 per cent more for it, and there's often no way of knowing for how long it's been defrosted or from where it originated. Supermarkets are making big profits from this tactic, but you can often get the same fish for a fraction of its price in its frozen form. Legally, fish sold loose do not have to state whether it was farmed or wild, nor where it originated from.

Another problem is that chemicals such as cadmium, lead or mercury have entered the oceans due to decades of industrial mercury emissions and this has contaminated some fish and seafood, particularly bigger, longer-living deep-water species such as yellowfin tuna, shark, marlin, halibut or swordfish. The data on the real dangers of fish mercury poisoning in humans is largely circumstantial and the results of studies are inconclusive, meaning it's hard to know what a safe level of fish consumption is.[26] The risks of heavy metal contamination in fish have been overplayed, but if you do eat a lot of fish and you're pregnant, it could become an issue. If pregnant women were to meet the dietary guidelines for fish and seafood, and chose predominantly high mercury-containing fish, they run the risk of eating too much mercury. There is some limited evidence to suggest that children born to women with slightly elevated mercury levels may develop measurable brain and nervous system deficits and are more likely than average to be diagnosed with attention deficit-hyperactivity disorder.[27] The evidence for any harm in the rest of us, however, is still unclear.

Fish contamination with plastic microparticles is an emerging concern and potential health problem. When surveyed in 2018, 73 per cent of 233 deep-living Atlantic fish contained significant levels of plastic. Starting at the bottom of the food chain, deep-sea marine organisms like plankton feed small fish such as sardines. These small fish are eaten by larger fish such as tuna, meaning that any contaminants consumed by the smaller fish progress up the food chain. This is coming from years of plastic pollution, which we dump into our oceans and rivers, perpetuated by our obsession with using plastic bottles for water. Similarly, when you eat mussels, clams and oysters you are eating deep-sea creatures that naturally filter our water. Any sediment (i.e., microplastics) which they can't break down is stored in their guts, and since you eat the entire animal, you will be ingesting the microplastic.[28] Belgians eat the most molluscs per capita of any country, with *moules frites* (mussels and chips) being the national dish. Because of this, Belgians may be ingesting 11,000 plastic microparticles per year. At the moment, most countries have more modest intakes, but this is likely to get worse as our obsession with using plastic continues. One estimate suggested that by 2050 there will be more plastic than fish swimming in our oceans. There are now microplastics found in the air and our food. We know virtually nothing about the potential risks of humans accumulating plastic in our intestines, and how our gut microbes will respond, but it's unlikely to be good news.

Fish is tasty and nutritious and probably isn't harmful for you and, if this chapter hasn't put you off entirely, can form part of a healthy and balanced diet. But saying we should all have it and can't live without it is crazy, as there's little (if any) strong evidence to support its supposed health benefits. For years we've been led by the government and food industries to believe that fish and fish oil supplements are good for our health. But the hard data on the benefits of fish eating is disappointing and it is hard to find good evidence for the over-hyped fish oil supplements. To date, there is no solid evidence to support the use of fish oil supplements in reducing risk of heart disease, and government guidelines have – unusually – changed to reflect this realisation.[29] Our demand for fish is becoming unsustainable, and if everyone were to follow government guidelines and eat fish at least once per week, we'd run out of stock and damage our precious oceans and their ecological systems and our planet even further.

By all means continue to enjoy fish, but pay more for better quality, seek better knowledge of the origin (farmed or wild) and consider it a treat rather than an everyday food. Generally speaking, the closer the fish is to the source of its original nutrients, the healthier they are for us, so it is better to eat small fish that eat plankton (such as sardines and herrings) rather than eating the fish that feed off them (such as salmon and mackerel). Eat more chia seeds, walnuts, flaxseeds and algae, as these plant sources will provide you with plenty of essential omega-3 fatty acids without destroying our oceans. If you like eating fish, try to vary species, and where possible choose fish that are sustainable, high in nutrients and, if you're pregnant, low in mercury. Picking sustainable fish is tough but you can look for labels or websites like the blue Marine Stewardship Council (MSC) globally, which, though far from perfect, indicates that your fish is wild, traceable and sustainable, or RSPCA Assured label (in the UK), which makes it easy to recognise products from animals that have led a better life. There are plenty of other organisations such as Friend of the Sea, Fish Wise and Global Fishing Watch who have been criticised but are working globally to help seafood businesses to become more environmentally friendly.[30]

For the moment, I will continue to enjoy a nice piece of high-quality sustainable fish meat once a week, but I know it won't save my life.

11.
VEGANMANIA

Myth: Veganism is the healthiest diet

Veganism no longer means eating limp lettuce leaves, tasteless tofu and a few bland beans. Nowadays, you're inundated with jackfruit panino, vegan Mac 'n' cheese, 'bleeding' beetroot burgers and vegan Kentucky fried chicken. The food industry has concocted plant versions of our favourite meats, cheeses and ice creams: Ben & Jerry's and Häagen-Dazs both offer vegan ice cream and America's largest meat producer, Tyson Foods, has started to reinvent itself as a protein company with meat substitutes as dairy and meat sales are declining. The global market for vegetarian/vegan products was worth $51 billion back in 2016, and that's set to increase to $140 billion within ten years.

Plant-based eating has become hipster and the number of vegans in the UK almost quadrupled between 2014 and 2019, with one in eight Britons identifying as vegetarian or vegan, and the US seeing a sixfold increase in vegans between 2014 and 2017, bringing the total to almost 20 million. Even those of us who aren't prepared to give up our steak or bacon any time soon are reducing our intake of animal products, with a third of British consumers now having meat-free days and one in three of us regularly buying plant 'milks'. Millennials are the ones who are driving the change and the most likely to shun eggs, dairy, honey and meat in favour of plant-based alternatives, with nearly half of UK vegans aged between fifteen and thirty-four. In 2020, Veganuary in the UK saw many brands offering vegan alternatives, including KFC, Burger King, Greggs and Pizza Hut. Many global food trends start in the UK, and the industry is watching closely. Vegans say that eating a plant-based diet will stop animal suffering, save the environment, improve our health and even add years onto our life expectancy. So is veganism the Holy Grail for us and the planet?

Many people claim simply to feel better and have more energy on a plant-based diet. Some people will benefit merely because they begin to think more carefully about what they are eating, and choose healthier foods and avoid random snacking. The benefits felt will depend on your initial diet. I certainly noticed this 'honeymoon' effect myself when I experimented with a vegan diet. If you're switching from a diet high in refined carbohydrates, processed meats and sweets to grains, fruits and veggies, you'll certainly feel better. And if you believe a food will make you feel better, the chances are it will, at least short term, which is basically the placebo effect. It could also be due to changes in the gut microbiome from eating a diverse range of plant foods.[1]

Many studies have looked at the effects of a plant-based diet on health and longevity, but results are mixed. A large meta-analysis of forty studies, which included over 12,500 vegans and 180,000 omnivores, concluded that a vegan diet is associated with favourable effects on risk factors compared with meat eaters.[2] Similar findings were seen in another review which suggested a plant-based diet could reduce risk of coronary heart disease (CHD) by up to 40 per cent.[3] But not all plant-based diets appear to be beneficial for heart health. A large follow-up study of 126,000 adults conducted over nearly thirty years found that while a high intake of healthy plant foods (wholegrains, fruits/vegetables, nuts/legumes, oils, tea/coffee) was associated with substantially lower CHD risk, less healthy plant foods (juices/sweetened beverages, refined grains, potatoes/fries, sweets) was associated with higher CHD risk. So, if vegans make healthy diet choices, they might have a reduced risk of heart disease. But does that mean they live significantly longer?

A study of 95,000 Seventh-Day Adventists found that being vegetarian was associated with a 12 per cent reduced risk of death from any cause compared with meat eaters.[4] This was a cohort study, which means that we cannot conclude that the vegetarian diet was the reason for the reduced risk of death; it could be due to other factors (confounders) such as physical activity. Importantly, there were a few key limitations to this study. Firstly, like many other vegetarian studies, it was conducted in a select sample of Seventh-Day Adventists: a religious group who are encouraged to avoid smoking and alcohol, and have been shown to lead healthy lifestyles with greater life expectancy compared to the general population. Secondly, the study had a relatively short follow-up time of six years; this doesn't allow much time to study diet effects on risk of death.[5] Finally, researchers included people who ate meat and fish (no more than once per week) in the vegetarian category. So the study may not be representative of what would be observed in other sample populations. A more recent and larger study followed a quarter of a million people over six years and found that while vegetarians had healthier lifestyles than omnivores there was no difference in death rates between both groups.[6] This is supported by a UK study which analysed 5,200 deaths and found that vegetarians and non-vegetarians had similar death rates.[7] Other studies, while showing health benefits and slight reductions in cancer, fail to find consistent effects.[8] Put simply, vegetarians and vegans are difficult to classify but do not appear to have a lower risk of early death compared with meat eaters.

But can a plant-based diet help us to solve the growing obesity epidemic? A study of over 60,000 Seventh-Day Adventists did find that vegans had the healthiest and lowest BMIs (23.6 kg/m2) out of vegetarians and meat eaters, suggesting that it may protect against obesity. However, as mentioned above, Seventh-Day Adventists are not necessarily representative of the US population. A small-scale clinical trial with sixty-two overweight women demonstrated that a vegan diet resulted in more weight loss (by about 3 kg) compared with a low-fat diet at one and two years' follow-up.[9] Another meta-analysis in 2016 of 1,000 dieters found that a vegetarian diet is superior to an energy-restricted diet by a small amount (2 kg), and that the biggest weight loss (2.5 kg) was seen in the vegan dieters. But the extreme vegan eating patterns are unrealistic and unsustainable for a lot of people, particularly since we know that most dieters regain all the weight lost within several years.[10] Interestingly, some vegan dieters go the other way, and develop unhealthy and obsessive thoughts about healthy eating.[11]

As part of our TwinsUK study, we analysed 122 British identical twin pairs who differed in their meat-eating habits, with one twin being vegetarian/vegan and the other a meat eater. Remarkably, there was only a small difference in body weight within the twins. The vegetarian/vegan twin was slightly slimmer by an average of 1.3 kg. Much larger differences of 4–5 kg between vegetarians and non-vegetarians were observed in the Adventist studies, yet those ignored the influence of genes. Our study suggests that genes, not just our dietary choices, also play an important role in determining body weight.

Many of the health benefits associated with vegetarian and vegan diets are probably related to eating a greater volume and variety of plants. Vegans certainly do eat a lot more fibre (the non-digestible parts of plant foods) compared to meat eaters. A systematic review found that eating higher amounts of fibre was associated with a reduced risk of cardiovascular disease, type 2 diabetes, colorectal and breast cancer, with the greatest benefits seen at intakes of 25–29 g per day, which is about double recommended levels in the UK and USA.[12] Eating more plants also means you'll consume more antioxidants, such as the anthocyanins found in brightly coloured berries, which are thought to improve gut health and protect against certain conditions such as heart disease and dementia.[13]

Avoiding dairy is potentially a problem for healthy vegans, however. For years it was assumed that since cow's milk is rich in calcium, the more you drink, the stronger your bones will be. The dairy industry, with help from government organisations, have poured millions into marketing this message, but this notion that dairy is always healthier is no longer supported by science. A substantial systematic review found there to be no evidence that increasing dietary calcium intake prevents fracture risk.[14] While some calcium is needed for strong bones, we need much less than previously thought, and most of us can get enough from loading up on vegetables (pak choi, broccoli) and other foods (tofu, nuts and seeds).

One estimate suggests that if everyone in the world was vegan and switched from dairy to soy milk, it would save half a billion hectares of land (about the size of Brazil), a billion tonnes of greenhouse gases and the same amount of water as if everyone stopped having showers and baths for a year.[15] One in three people in the UK now buy some plant-based milks, and even the dairy giant Danone have invested $60 million in vegan milk production. While the craze for plant-based milks has taken off, from soy, almond and oat to hemp 'milk', there is now a dispute in many countries about whether they should even be called milk, since nuts and beans obviously don't lactate. If you do decide to switch to white-coloured plant juice (fake milks), it's hard to know which one to choose. While they all have much lower impacts on the environment compared to dairy milks, they each come with their own issues. Almond milk requires irrigation of trees in deserts, which places huge pressure on water resources; rice milk produces methane gas released by the bacteria in the flooded rice paddies; and soy and oat milks use more land, which means replacing trees. Non-dairy alternatives have come a long way in the last decade. In addition to being more environmentally friendly than dairy milk, nutrient profiles have improved and many non-dairy milks now contain equivalent calcium levels to cow's milk.[16]

Evolving to eat meat over 2 million years ago has been our evolutionary key to success. So can you get enough nutrients without eating meat? Contrary to popular belief, most healthy people in developed countries consume more than enough protein from their diet.[17] Although vegans and vegetarians do consume about a third less protein on average per day compared with omnivores, they still exceed the daily recommended amount of protein.[18] Tofu, legumes and pulses, certain grains, nuts, seeds and mushrooms are the most common source of protein for vegans. Another myth is that vegans lack essential amino acids. While this may be true for those on poor diets, a balanced diverse vegan diet has consistently been shown to be nutritionally adequate for essential amino acids.[19] There are some real problems, though. One is the increased risk of nutrient deficiencies in vitamin B12 and iron, which are harder to come by in plants and grains. Vitamin B12 deficiency (characterised by tiredness, mood swings, tingling in arms and legs, and a sore tongue) is common, and as a result many vegans resort to taking lots of nutritional supplements.[20] Relying on artificial supplements is for me not an indicator of a balanced and healthy diet. Moreover, despite taking supplements, many people remain with low blood levels, partly because some people need more than others due to genetic factors. And finally, vegans also have lower iron stores – although men are more likely to be affected.[21] The lack of iron stores increases the risk of iron-deficiency anaemia but the flip side is that high iron levels may be associated with diabetes and heart disease.

A bigger concern is the rise in veganism among children and even pets. If you raise a cat on a vegan diet you will probably kill it. Dogs, on the other hand, are omnivores and can in theory survive on a vegan diet. But what about the increasing numbers of vegan children? While it is possible to raise them to be healthy, it isn't easy and there are serious health consequences of getting it wrong. Studies have shown that children raised on a vegan diet are often smaller and have low levels of certain nutrients such as riboflavin and B12, which, when extreme, has led to high-profile deaths.[22] In France, raising your children as vegan is classed as criminal neglect. As vegan-ism becomes more popular in adolescents, it often overlaps with other exclusion diets like gluten-free diets that can lead to modern eating disorders, such as orthorexia, an abnormal fear of eating unhealthy food.

Contrary to popular belief, not all vegans are super-healthy and live on a diverse leafy diet. Many eat chips, biscuits, cakes and processed vegan versions of meat and cheese, which are often pumped full of chemicals, sugar and saturated fats. Gregg's, the UK high-street bakery chain, recently launched a vegan sausage roll which is now a best-seller, and Burger King offers a vegan burger. Regardless of whether your sausage roll is from a pig or Quorn, or your burger from beef or soy and fun-gus, it doesn't make one type healthier than the other; they're both highly processed and high in calories, saturated fat and salt. Vegan and vegetarian processed foods are marketed as healthier than they actually are, and some, like vegan fish fin-gers, contain up to forty artificial ingredients.

So, veganism per se is not necessarily healthy. Most of the benefit is probably just through eating a greater variety of plants and fibre, which can still be achieved by people eat-ing small amounts of meat and dairy. Don't feel pressured into buying these plant-based alternative products, which are often pumped full of additives, sugar and fat and could do you more harm than good. If you're like me and you love a ripe, oozing raw-milk Brie or an occasional grass-fed organic piece of meat, there's no need to forgo these pleasures entirely. You can still be healthy by eating all forms of plants, grains, nuts and seeds, drinking less milk and occasionally eating meat and fish, opt-ing for high-quality and less highly processed foods. While there is no doubt avoiding meat and dairy delivers tremendous benefits for the environment, as land is used directly for plants rather than inefficiently for animal feed, strict veganism is too much for many people. Consider becoming a part-time vegan or flexitarian, reduce your meat and dairy consumption and replace them with real plants to reduce global warming. If everyone started by cutting out meat just one day a week, we would all rapidly reap the benefits.

12.
MORE THAN A PINCH OF SALT

Myth: We all need to reduce our salt intake

We've been warned about the dangers of salt for some time. Epidemiologists have highlighted the potential problem since the 1980s and, for the last twenty years, governments have been encouraging us to reduce our intakes through nation-wide campaigns, salt taxes, food labelling and educational initiatives. We are told that reducing our daily salt intake below 6 g (one and a quarter teaspoons) is the key to reducing blood pressure, strokes and heart disease, and will save the US economy up to $32 billion each year due to potentially reduced health costs (although people who live longer also incur other health costs).[1] Reducing salt in our diets has been one of the UK Department of Health's top priorities for over a decade.

Average intakes of salt have decreased by about 14 per cent in the UK since salt-lowering initiatives began in 2001, and by 23 per cent in Japan, mainly by encouraging people to use less soy sauce.[2] By 2010, the salt limit of 6 g per day was ingrained in American nutritional guidelines (where it is called sodium) and they pushed it even lower to below 3.8 g (just over half a teaspoon) for everyone of African-American heritage or those with high blood pressure, heart failure, kidney disease or diabetes, which is nearly half the adult population. The stakes have been raised again recently and international bodies such as the World Health Organization and American Heart Association have set in 2018 even more ambitious goals to reduce all our daily intakes to below 5 g of salt or just one teaspoon per day.

The average intakes of salt vary across countries but most of us consume twice this amount, between 9 and 12 g per day. In the US and the UK, salt intakes have remained consistently nearly double the recommended guidelines for the past decade. Several years ago, I strongly believed that we were all eating too much salt and needed to cut back. But I may have been misled.

Common table salt is mainly composed of two minerals, sodium (40 per cent) and chloride (60 per cent), which both have important functions in the body, for muscles, nerves and fluid balance. The reason that salt is added to food in every country in the world is simple – it tastes better. It accentuates flavour, balances out dishes, mellows bitterness, and every good chef will tell you that the first essential skill is how to properly salt your food – and the biggest crime is undersalting. When the Hadza hunter-gatherers have extra honey to trade, their first thought is to swap it for salt. Salt has been a precious commodity throughout human history, and Roman legionnaires were paid with it. Adding salt is an essential part of the preparation of traditional fermented foods such as kimchi, pickles, sauerkraut, and many cheeses which have been eaten for centuries. High amounts of salt prevent the growth of harmful bacteria, which can cause foods to go bad. Salt is required for human survival and has a key role in enhancing flavour and preserving food. So why are we now so concerned about it?

The link between salt and blood pressure has been postulated for centuries and seems logical, as adding salt to water increases its pressure. In the 1990s, a series of observational studies showed that salt levels in the diet mirrored blood pressure, and when people with low salt intakes and low blood pressure migrated to new locations with high intakes of salt, their risk of high blood pressure increased. Until recently, I was impressed by the data. The evidence was so compelling that it was not just a question of if, but more a question of how drastic the reduction needed to be.

We get 70 per cent of our salt from processed or restaurant foods – not the salt shaker. British food manufacturers agreed to voluntary salt-reduction targets without a struggle, presumably so that they could reformulate cheaply and market their products as 'low salt' and thus appear to be healthier. The industry started to reduce salt levels in many processed foods, under the eyes of the Food Standards Agency, although this initial cooperation fizzled out after 2010 when the Department of Health took over. By 2019, nearly half of the agreed voluntary salt targets were not being met. There are strong anti-salt advocacy groups in many countries who correctly (and aggressively) call out highly processed foods for their high levels of hidden salt. These examples include many breakfast cereals and mass-produced Cornish pasties that contain as much salt as seven portions of salted peanuts. A 2013 British survey of 700 popular restaurant dishes found many that contained more than an entire day's recommended salt intake.[3] Customers in fast-food restaurants regularly under-estimate their salt intakes by sixfold when asked,[4] and many of us would be surprised to find that our favourite muffin, doughnut or bagel is laden with salt to enhance its sweetness and extend its shelf life.

Around the world, official salt-reduction strategies include working with industry to reformulate food products, establishing sodium or salt targets for foods, educating consumers, revising labelling schemes on food packages, and in some cases taxing high-salt foods. Twelve countries have reported reductions in intake levels thanks to their policies and these get good publicity, though we don't tend to hear about others that have failed. South-East Asian countries such as Bangladesh, Thailand and Indonesia enjoy salty food, but have so far ignored the global pressure to adopt voluntary or mandatory salt target levels. The salt police predict these countries will experience higher levels of heart disease and hypertension in years to come.

Observational studies and clinical trials have shown that reducing salt intake in people with hypertension (high blood pressure) can help to lower their blood pressure a little. What we are not told is that any changes from reducing salt intake pale into insignificance next to the proven benefits of blood-pressure medications. Anti-salt lobby groups, diet professionals and governments want us to believe that any health improvements are the result of their public health interventions rather than drugs.

There is also growing evidence that some people respond much more strongly to salt than others, and are known as salt sensitive. The relatively new concept has caused controversy as to whether salt sensitivity is a separate subgroup or disease, or part of the normal spectrum of how we respond to food. The food and pharmaceutical industries aren't keen to promote this idea as they worry it will reduce demand for their products and require warning labels on foods. If you have African ancestry, you are more likely to be sensitive than Europeans or Asians, on average, but within every group there are big differences on a continuous scale. I am currently part of a UK consortium (AimHy) looking to see if we can predict from genes, blood tests, gut microbes and ancestry who, from groups of Europeans, Asians and Africans, will respond best to a range of blood-pressure drugs. We have shown from our own twin studies that genes play a major role in influencing blood pressure, and twin studies over thirty years ago showed that blood-pressure response to high-salt diets is highly variable between people and is influenced by genetics.[5] Several studies of Europeans and Asians have shown having one or more variants of a common gene considerably increases your risk of salt sensitivity.[6] But, as so often, we have been prescribed oversimplified, one-size-fits-all advice which fails to reflect the real science.

For most healthy people, the reduction in blood pressure as a result of salt reduction is surprisingly small and clinically trivial. A review of thirty-four studies found that for people with normal blood pressure, reducing their salt intake to one and a quarter teaspoons per day resulted in a very small reduction in systolic blood pressure of 2.4 mmHg and diastolic blood pressure by about 1 mmHg (only around 1–2 per cent).[7] This may make you question whether a life following a salt-free, tasteless diet is really worth it.

But even if there was compelling evidence that low salt intakes led to long-term reductions in blood pressure, this would only be significant if this reduced risks of cardiovascular disease and death. In fact, contrary to popular belief, studies of salt reduction have not found any reduced risk of heart attacks, strokes and death. In 2014, an independent review analysed eight studies of 7,284 participants.[8] While diet advice and salt substitution reduced the amount of salt eaten, which led to a small decrease in blood pressure over six months, this did not result in any significant benefits on heart attacks, strokes or death. Only one salt-reduction intervention study back in 2006 has shown beneficial effects – it may be no coincidence that it was also funded by a salt-replacement company.[9] So, although we lack long-term studies lasting more than ten years, it appears that for most people salt reduction has few clear benefits.

But it doesn't end there. Another shock came recently when randomised clinical trials of patients with diabetes on low-salt diets reported that, rather than getting better, patients were consistently dying earlier. One study, following 638 diabetic patients, found that those with low salt intakes had an increased risk of death.[10] Several small-scale clinical trials have now shown that a low-salt diet may reduce the body's response to insulin, because of a protective response and release of chemicals such as the stress hormones (like adrenaline) from the kidneys and an increase in blood fats.[11] More high-quality clinical trials are required to confirm this, but current evidence suggests that telling diabetics to reduce their salt intake to low levels could actually be harming them. These findings are important for many of us, as diabetic or pre-diabetic individuals (like me) now make up a large proportion of our population.

A study in 2018 caused a stir in the black-and-white world of public health strategy. The large population-based study observed 95,757 people from eighteen developing, mainly Asian, countries over eight years. It found that those in the top third of intakes of salt had increased heart problems and stroke, as expected, but there was no risk at intakes below 12.7 g per day, although this is still way above average US and European levels. Those who consumed the lowest amount of salt (below 11.1 g/day), instead of being protected, actually had greater risks of disease.[12] The sweet spot of health was somewhere in the middle. Those consuming the most salt came from China and these high-risk levels are only seen in 5 per cent of populations in the West. This should have led to us rethinking public health policies of reducing everyone's intake to below 5 grams of salt. Instead the anti-salt lobby attacked the study as flawed and said the case against salt had been proven.[13]

As with so many of the myths in this book, it's the fear of salt that is potentially more harmful than the salt itself, and the measures we and the food industry take to avoid it may be more dangerous for our health. These fears about salt have led to many food companies adding in lots of other chemicals such as potassium, MSG and lysine so they can label their products as 'low salt'. While potassium supplementation has been linked to small reductions in blood pressure, you could probably get similar or greater benefits from eating fresh fruits, vegetables and wholegrains with no risk.[14] A few years ago, worried about my salt consumption, I swapped my regular table salt for LoSalt, which contains a higher percentage of potassium chloride instead of sodium chloride. It tastes similar to salt but leaves a funny metallic chemical taste. Too much potassium can be dangerous for people with conditions like heart disease, liver disease and diabetes as these are already associated with raised blood potassium levels, and the additional chemicals in the salt replacement could push you over the edge. Chemicals such as potassium can also interact with common medications (including blood-pressure tablets, like diuretics and ACE inhibitors). Kidney doctors have gone as far as calling potassium salt replacements 'deadly' because they can be fatal for kidney-disease patients on dialysis, as they trigger acute heart problems.[15] Although food companies and the USDA say they are safe, much less is known about the effects of adding lysine and MSG to our food. Observational studies in Asia, where use is ten times more than in the West, have suggested that excess MSG may be linked with obesity and metabolic syndrome.[16] We know little about lysine in humans, but it is widely used in animal feeds and studies in rats show it increases growth and body size.[17] These examples show the potential health risks of tinkering with something as simple as salt, by adding chemicals and mixtures that we know next to nothing about.

It is hard to avoid the conclusion that the current guidelines about salt are flawed. People with either very low or very high intakes of salt seem to have higher mortality. There's no doubt that very high salt intakes are linked to higher rates of hypertension and heart disease, but much of this can be explained by people eating too much salty ultra-processed food which also has other problems. While there are likely to be benefits in reducing salt for people at the maximum levels of intake, it is unlikely to have any significant effect on the health of the majority. Despite this, dietary advice hasn't kept up, and most guidelines and nutritionists are still telling people they need to reduce their salt intake or risk a lifetime of heart disease, strokes and hypertension. As so often in nutrition, the pressure groups have once again focused on one single aspect of our diet, this has led indirectly to reformulation of processed food, without looking at the more important factor of overall quality. Reducing dietary salt intake in the whole population clearly isn't working and there's emerging evidence to suggest it could actually be harmful for some.

As we've seen from successive gold-standard independent reviews, for most individuals, any benefit of salt reduction on its own is trivial in terms of cardiovascular health. This is starting to sound worryingly like the cholesterol/saturated fat story of the 1980s, driven by a small group of salt-reducing zealots. Again, we've failed to look at the bigger picture and the complexity of our food interactions. Salt response is highly individual, with certain ethnicities and races being more sensitive to it than others. This is another reason why campaigns and guidelines should not lump us all into the same category. It is a recurring theme: governments avoid telling us about variations in our individual risk as they are so focused on their own political, financial or box-ticking agendas.

While there are some clear exceptions, most people should be able to enjoy salt as part of a high-quality, balanced diet and if you don't eat junk food every day, you probably don't need to worry about adding salt to cook your pasta, improve the tenderness of your meat or enhance your tomato salad. Of course, getting your salt fix from artisanal breads, cured meats and cheeses is preferable to ultra-processed burgers, burritos, pizzas and crisps. But if we avoid these highly processed excesses, most of us can still enjoy our food without guilt, and should take current strict guidelines with a reasonably large, and surprisingly healthy, pinch of salt.

13.
COFFEE CAN SAVE YOUR LIFE

Myth: Drinking coffee is bad for our health

Most of us rely on caffeine to get us going in the morning. This psychoactive stimulant is commonly consumed in the form of tea or coffee, which are still the world's most popular drinks. Americans consume 400 million cups of coffee per day, making them the leading consumers worldwide. The US coffee market is worth $18 billion annually, and high-end speciality coffee sales are increasing by 20 per cent per year. The UK isn't far behind, with Britons drinking almost 100 million cups per day and it has even overtaken tea drinking. Caffeine is best known for its stimulatory effects on the brain – it helps us to stay awake and feel alert. More recently, it is being added to foods, drinks and even diet supplements. But many people say caffeine is as evil as alcohol. Coffee has taken most of the blame, probably because of its higher caffeine levels in the finished product than tea or chocolate. We've been told that coffee is linked with poor sleep, heart disease and even cancer. So is caffeine really that dangerous?

Doctors used to say that too much coffee was bad for you due to the caffeine. Before the 2000s, many case-control observational studies (which are prone to bias) looked at past coffee consumption in sick people compared with healthy controls. There was a positive association with heart disease risk in those who drank lots of coffee.[1] This led to scientists conducting studies in rats. High doses of caffeine sped up the heart rate and sometimes caused rat arrhythmias (abnormal heart beats) and even certain cancers. For years this was the accepted norm until a more recent systematic review of human studies concluded that caffeine does not have a significant effect on arrhythmias.[2] One analysis of thirty-six studies found that moderate amounts of coffee (three and a half cups per day) actually reduced the risk of heart disease, and even heavy coffee consumption was not associated with elevated risk of heart disease.[3] Yet another review combined twenty-one prospective studies looking at coffee habits in over a million people from Europe, the US and Japan.[4] Moderate coffee drinking (three to four cups per day) was associated with reduced risk of death by 8 per cent and heart disease by 20 per cent. We should be wary of the limitations of this kind of data, but forcing people to drink high doses of coffee in a trial could be tricky, so these estimates are probably as good as we will get.

Scare stories often appear in newspapers about acrylamide, which is produced in small quantities in coffee when the beans are roasted. In high amounts it's been linked to cancer in rodents and in 2018, coffee shops in California were so worried about legal action that their drinks now come with a warning about it.[5] Like hundreds of other commonly consumed chemicals, it has been classified by the World Health Organization as a carcinogen that could potentially cause cancer if consumed in large amounts. The media love scary food headlines; I've already mentioned the hype around similar stories of overcooking meat and burnt toast causing cancer due to acrylamide.[6] If you were to unpick the hundreds of chemicals found in foods and investigate them in isolation, you'll almost certainly find one that is considered to be harmful to rodents in massive doses (and you could start your own WHO lab). But the relevance of this research to humans is weak. Nearly forty years ago as a medical student I wrote a paper based on global data suggesting that coffee could cause cancer; it certainly helped my career, but in retrospect, it didn't help science.

Other common concerns relate to its effects on our toilet habits. Caffeine stimulates the bladder to produce urine more quickly than usual.[7] You might need to rush to the loo more often, but there's no evidence to suggest it actually dehydrates you. Caffeine is a powerful chemical and some individuals are naturally more sensitive. For example, it acts as a trigger for some people with irritable bowel syndrome (IBS), for whom it may give cramps and diarrhoea. It can also keep you awake at night. Caffeine acts by blocking the normal actions of a brain-relaxing chemical called adenosine, which normally makes you sleepy. By blocking adenosine, caffeine increases alertness and concentration, which may explain why caffeine reduces the risk or delays the onset of Alzheimer's and Parkinson's disease and why elite athletes perform marginally better on it.[8,9] On average, caffeine levels increase in the blood thirty minutes after drinking a coffee, peaking at two hours, and disappearing after four to seven hours, when the chemical is disposed of by the liver. The effects of caffeine drunk before six in the evening are likely to have worn off by the time you go to bed, but its metabolism is highly variable in different people. Caffeine in small doses can disrupt sleep in sensitive people, and if you have insomnia or struggle to sleep well, you could switch to decaf or just stop drinking caffeine in the early afternoon.[10]

Some people with mental health conditions are afraid that caffeine will make their symptoms worse. Consuming too much caffeine can cause jitteriness and anxiety, which overlap with some psychiatric disorders.[11] For this reason, many inpatient psychiatric facilities ban caffeinated beverages. But the research is contradictory, as caffeine consumption in some studies shows a preventative effect. A follow-up study of 50,000 middle-aged US women found that people who drank the most coffee had a 20 per cent reduced risk of subsequent depression.[12] Another review of three studies analysed data from 47,000 participants and found that, strangely, those who drank four or more cups of coffee per day were half as likely to commit suicide.[13]

Researchers cannot yet distinguish what component of coffee might be beneficial for our health, as the caffeine itself may not be crucial. Coffee contains high levels of the antioxidant chemicals, polyphenols, which are likely to be beneficial due to their role in feeding our microbes.[14] Reassuringly, the roasting process doesn't destroy them; in most cases the polyphenols and antioxidant power actually increases. Polyphenols aren't the only beneficial component in coffee – a mug of coffee is a reasonable source of fibre, with each cup having around half a gram. So drinking a few cups throughout the day gives you the same amount as eating a bowl of cereal or small banana. The fibre is fermented by the microbes in your colon to produce helpful short-chain fatty acids, which help other beneficial bacterial species in our guts to flourish.[15] So thanks to the fibre and polyphenols, coffee wakes up our microbes as well as our brains in the morning.

Even decaf coffee has an impressive polyphenol content. It's usually made by rinsing coffee beans with chemical solvents. The process removes most, but not all the caffeine; typically around 97–99 per cent on average, but levels can vary. The newer methods conserve more antioxidants, so you can still get your daily dose of polyphenols from decaf coffee. Blind tastings have shown that it's difficult to differentiate between caffeinated and decaf coffee. Most normal people drinking decaf coffee can't be fooled into experiencing symptoms typically associated with caffeine, such as increased alertness. But some anxious people can still develop anxiety symptoms when given decaf coffee.[16] Another study found that if you think you're still having caffeine, unpleasant withdrawal symptoms from caffeine abstinence are reduced, showing many of us can be easily tricked.[17]

Most Western guidelines err on the side of caution and say up to 400 mg of caffeine per day (four cups of instant coffee or three filter coffees) is fine.[18] This is considered to be a safe level of caffeine that most healthy adults could consume on a daily basis without experiencing negative side effects. There's limited research in children and adolescents, but EFSA (European Food Safety Authority) say that 3 mg per kilo of body weight per day is safe for children. So a fourteen-year-old who weighs 50 kg could safely consume 150 mg of caffeine per day, which is the same as a small filter coffee. The guidance for pregnant women is less clear-cut and many women avoid caffeine altogether (see chapter 14, page 147), even though 200 mg per day (two cups of instant coffee) is considered safe. Caffeine tolerance is highly personal, and there are some factors you can't control – such as your genes. Our twin studies have shown that taste and food enzyme genes can influence your personal preferences for how much you like strong bitter flavours like coffee.[19] Our latest PREDICT study shows that coffee drinkers have very different gut microbes, and these could also play a part in tolerance levels.

Many drugs interfere with how rapidly caffeine is metabolised before it affects the brain. A smoker with nicotine in their blood will need twice as much coffee to get the same hit as a non-smoker. Hormones also play a role, with women being more sensitive to caffeine than men, which is increased further by contraceptive pills or antidepressants, meaning even small doses can keep you awake. Alcohol also increases the effects of caffeine, exacerbating sleep problems, and vegetable lovers who eat broccoli and other crucifers regularly will need more coffee as some of the polyphenols reduce the effects of caffeine. So if you're female, on the Pill and a non-smoker who dislikes kale, you should ignore the guidelines and may be best sticking with decaf in the evenings.

Food and drink companies add large amounts of caffeine to their products for its additional health benefits. Virtually all sports drinks, energy bars, weight-loss supplements and diet drinks have added caffeine along with a string of health claims. A can of Red Bull contains the same amount of caffeine as two espressos, while drinks like Relentless and Monster can contain double that amount. Some products claim that the caffeine boosts metabolism, accelerates weight loss and improves sports performance. The effects, if any, are marginal, with some small studies suggesting that while increased rest-ing metabolism may possibly burn an extra 70 calories, it won't counteract the added sugar.[20] Others suggest it improves exer-cise performance, but this is only by 1 per cent or a few sec-onds, so useful only for professional athletes.[21] The effects are likely short term and probably won't make you skinny or turn you into the next Usain Bolt. While moderate amounts of natu-ral caffeine are considered safe, much less is known about the effects of synthetic caffeine on our health and our microbes, and therefore I'd recommend steering clear of any food or drink product with this additive, as they usually have added sugar and a dozen other chemicals in them, which our bod-ies can begin to crave more of. Stick to high-quality coffee or tea, which contains natural caffeine and far fewer ingredients. But not all teas or coffees are healthy; cream frappuccinos are ultra-processed and sometimes contain over 700 calories. You need to be aware that the amount of the drug (caffeine) that you get in each cup also varies widely. This depends on the type of bean, the roasting process (lighter roasts have more caffeine), types of coffee and serving size, as well as the barista. Filter coffee has around 140 mg per cup, and a mug of instant coffee has around 80–100 mg, while a shot of espresso var-ies most and has between 40–200 mg. As I mentioned above, decaf isn't entirely caffeine free and contains around 3 mg per cup, which some people can still be affected by.

It's clear that caffeine is no longer considered 'deadly'. In moderate amounts, tea and coffee don't do us any harm and there's increasing evidence that they're good for us. Coffee isn't just about caffeine; it contains some fibre and, similar to tea and dark chocolate, it's packed with polyphenols, which are known to have gut health benefits, and should stay part of our daily diets. Everyone is different. We all have our own personal tolerance levels, and we should keep experimenting to find the right dosage that suits us best. Many of our best creative ideas in recent centuries have arisen from coffee shops, and for once we may have picked a drug which, if we get the dosage right, suits many of us.

14.

EATING FOR TWO

Myth: Pregnancy nutritional advice is reliable or evidence-based

Pregnancy is meant to be an exciting time but it is often overshadowed by anxiety about what you can and can't eat and whether you should be 'eating for two'. Many countries have devised their own pregnancy nutrition guidelines and recommendations, which supposedly steer women in the right direction. It all seemed eminently sensible to me, until I started speaking to pregnant women from different countries and realised that the advice given around the world is quite different. While researching for this book, I conducted a brief online survey with hundreds of nutritionists and pregnant women in eleven different countries to find out just how confusing and contradictory the advice is. Strangely, there are no up-to-date international pregnancy nutrition guidelines. This is a far cry from the long list of foods that UK and US guidelines advise against consuming (sushi, deli meats, raw eggs, alcohol, uncooked meat, soft cheese, unpasteurised milks, pâté, etc.).[1] Given their similarity, I'll group these together with those from Canada and Australasia and call them 'Western guidelines'. The first 1,000 days from conception are a critical window of development. Ensuring mothers are informed and have access to fresh nutritious foods in this window is crucial for the future health of their children, and for themselves. Pregnancy is the biggest risk factor for triggering obesity in women, and we should be helping women eat enough of the right foods, instead of scaring them from eating anything but a few processed foods.

Coffee and caffeine intakes are part of many national guidelines. High levels of caffeine consumption have been linked with having a lower birth-weight baby, which can have health consequences later in life.[2] Western guidelines therefore recommend limiting caffeine to a 'moderate' intake of 200 mg per day. This is equivalent to one real coffee per day or two mugs of instant coffee, or double the amount of tea. The nutritionists that I spoke to said that in the US they're encouraged to switch to decaf, whereas in Italy espressos, cappuccinos and macchiatos are fine. Herbal teas are even less clear-cut. The Japanese sip unlimited green tea during pregnancy, despite it containing moderate amounts of caffeine. Western mothers who I spoke with were afraid of drinking some teas due to fears of miscarriage, whereas East Asian women were actively encouraged to drink them by herbalists and doctors due to supposed health-promoting effects on the foetus. Some midwives around the world recommend raspberry leaf in the final weeks to trigger labour, as it's rumoured to stimulate the uterus, whereas others suggest avoiding fennel seed and liquorice tea for fear of miscarriage, but evidence is hard to find.[3] More worrying is that nearly a third of pregnant women in Western countries take some specific form of herbal medicine during pregnancy, and some of the hundreds tested are likely to be unsafe.[4] Western guidelines say that up to four cups of herbal tea per day is fine but you should vary the types consumed throughout pregnancy. This is a perfect example of the recommendations being based on guesswork with very different levels of caution depending on the prevailing culture.

Nearly everyone agrees that alcohol, which is a chemical neurotoxin that impacts development, during pregnancy is bad and can sometimes lead to foetal alcohol syndrome, with serious effects on a baby's brain and behaviour, but it is very uncommon, affecting less than 2 per cent of the babies of heavy drinkers.[5] A five-year follow-up study of 1,600 women and their children found that small amounts of alcohol consumed occasionally (as opposed to regularly) during pregnancy probably isn't harmful.[6] That's good news, because 10 per cent of women globally are estimated to drink alcohol during their pregnancy.[7] This is partly due to the fact that one in six pregnancies are unplanned and therefore lots of women have knocked back a few drinks while not knowing they were in the early stages of pregnancy. Even some of the nutritionists that I spoke to said that they, despite knowing the supposed risks, had enjoyed the occasional glass of wine throughout their pregnancies, so if you are sensible, the odd sip of wine or beer here and there isn't likely to do you or your baby any harm.

Weight gain during pregnancy is another common concern. A large systematic review of 1.3 million pregnancies from around the world found that nearly half of the women gained more weight than recommended, increasing the risk of bigger babies or Caesarean deliveries.[8] Perhaps this isn't surprising as several British mothers that I spoke to said that they were still being told by health professionals to 'eat for two'. This myth refuses to die, despite it being widely accepted that women only require an extra 200 calories (a small bowl of cereal or a large scoop of ice cream per day), but – importantly – *only* during the last three months of pregnancy.[9] Emerging evidence suggests that gaining too much or too little weight during pregnancy can also increase risk of high blood pressure, obesity and diabetes in offspring.[10] Some countries, including the US and France, regularly weigh women during pregnancy, but others frown upon this.

Routine weighing began in the UK in the 1940s due to concerns that wartime rationing would prevent pregnant women from getting adequate nutrition.[11] In the 1970s, the focus shifted and regular weighing was used to prevent excessive weight gain, because of health risks. But reviews in the 1990s led to a general consensus that routine weighing generated anxiety in pregnant women without convincing evidence it worked, and it was dropped.[12] Currently, most British women are only weighed at their first antenatal appointment (twelve to fourteen weeks) and have their bump measured throughout pregnancy instead. This caution has persisted and there are no official UK guidelines for recommended pregnancy weight gain; this is despite a massive obesity epidemic with dramatic health consequences in pregnancy.[13] The US and French weighing practices, though not perfect, do allow appropriate dietary and lifestyle advice to reduce health risks to both mother and baby. In the UK and other Western countries around half of all pregnant women are overweight or obese, and studies show interventions can help reduce obesity in the mother after birth.[14]

The mothers that I spoke to had mixed opinions about regular weighing. Some said that it provided them with re-assurance that their baby was growing adequately (particularly those with morning sickness), but one US mother said that her doctor nagged her about not gaining too much weight, despite being a normal weight beforehand, which caused her unnecessary stress. Another US mother was left feeling mortified when her doctor said to her, 'You're getting huge, how many are you growing in there?' So the overwhelming message was that women are happy to be weighed during pregnancy, and it doesn't have to be a taboo subject, but the interaction needs to be sensitive, and without judgement.

When it comes to foods which need to be avoided, there are huge differences between countries. Western nutrition guidelines recommend avoiding cold, cured meats like salami, chorizo and pepperoni or undercooked meats (i.e., a rare steak) due to a small risk of contamination with toxoplasmosis, which in a handful of cases has caused miscarriage or damaged the foetus.[15] Western guidelines suggest either avoiding these foods, cooking meats thoroughly or freezing cured meats for four days prior to eating to kill off any parasites.[16] A Russian dietician told me that when she asked her paediatrician whether she should avoid cured meats he laughed in her face, before saying, 'You think that pregnant women in Russia have the luxury to pick and choose what they eat?' Smoked meats and fish (which are cured but uncooked) are considered a traditional staple in the Russian diet regardless of pregnancy. A Portuguese woman told me that her doctor recently reassured her that smoking the odd cigarette here and there is fine during pregnancy, to avoid her baby becoming stressed. Western doctors could be struck off if they were overheard telling that to patients, so I think it's fair to say that there are some big cultural differences.

Nearly all the women I spoke to told me that they avoided raw fish and sushi throughout pregnancy to reduce risk of infections.[17] In contrast, most nutrition guidelines actively encourage fish consumption (particularly oily fish) throughout pregnancy, although the French say to avoid smoked salmon or trout.[18] Eating sushi during pregnancy is not recommended in most countries, but a Japanese dietician told me that raw fish isn't restricted at all, and the Japanese laugh at the notion that it should be avoided. The vast majority of sushi is thawed from frozen, killing any rare parasites. Even in fish-loving Japan, though, it's recognised that some types of fish should be limited during pregnancy due to their high mercury content.[19] Overly cautious Western guidelines recommend completely avoiding any fish that contains high levels of mercury, such as marlin, swordfish, shark and bluefin tuna. But the Japanese guidelines simply advise reducing intake of these fish to once or twice a week. The Japanese have been eating raw sushi (and probably a fair amount of mercury too) during pregnancy for years without any issues, so their advice seems sensible.

What about raw eggs? For years, pregnant Westerners have been avoiding fresh mayonnaise, poached eggs and mousses due to stories about raw eggs causing salmonella.[20] While there are a few case studies linking miscarriage to salmonella infections, only one in a thousand people probably ever experience a salmonella infection in their lifetime, so the chances of this happening to you, especially during the short nine months of pregnancy, are extremely small.[21] Again, this amuses the Japanese, who for years have been eating traditional probiotic dishes such as natto (fermented soybeans with a raw egg on top) throughout pregnancy without any issues. In the Philippines, women are actually encouraged to eat raw eggs prior to pregnancy to 'lubricate the birth canal'.[22] In the UK, previous recommendations to avoid raw and runny eggs have recently changed. Now, if the egg has a Red Lion stamp on it, it means it carries a low risk of salmonella and is safe to consume. In other countries like the US, antibiotics, chlorine or vaccines in poultry are widely used to nuke any bacteria, and untreated eggs need to show this on the packaging. Most commercial egg-containing products like mayonnaise, meringue and chocolate mousse contain pasteurised egg anyway, destroying harmful bacteria.

Dairy is another minefield. As if women don't have enough to worry about, Western guidelines say that all cheeses with soft white rind such as Brie and Camembert and soft blue cheeses should be avoided unless cooked. The reason being that these types of cheeses are less acidic than hard cheeses like Cheddar, and contain more moisture, making them an ideal environment for harmful bacteria like listeria to grow in. They also recommend completely avoiding raw-milk cheeses which are not pasteurised (i.e., local artisan cheeses) due to a small risk of toxoplasmosis. Even the French advise avoiding unpasteurised dairy, most soft or raw-milk cheeses and wash rind cheese. Apparently not all French women follow this advice, and several French nutritionists told me that they and their friends continued eating all cheeses throughout pregnancy, especially if they'd had toxoplasmosis in the past and were therefore immune to it. French, Austrian and Italian women can afford to be a little more relaxed with their dairy consumption as they are given routine blood tests at the beginning and throughout pregnancy to check for any new infections such as toxoplasmosis and receive immediate treatment if necessary.

The actual risk of infection with listeria is increased twentyfold in pregnancy, but absolute chances of contracting it are extremely low. To put this into perspective, listeriosis affects around twenty pregnant women per year in the UK and three-quarters of those have a normal baby.[23] Even if you were unfortunate enough to get listeriosis during pregnancy, it's highly treatable, and if caught early it rarely causes issues to the foetus.[24] There is estimated to be less than one death per day globally due to all listeriosis infections compared with a massive 3,300 deaths per day due to car accidents, so you're thousands of times more likely to die or harm your baby in the driving seat than you are from eating a hunk of Brie.[25] Women are not usually told to avoid bagged salad, watercress and vegetables, even though it's these, rather than cheese, that have been behind the most recent listeria outbreaks in the West.[26]

A lot of these precautionary recommendations are based on little, if any, scientific evidence, mostly observational and often contradictory. In Asian countries, many women restrict their diets due to ancient traditions and beliefs about certain hot (yang), cold (yin) or spicy foods causing miscarriage.[27] In traditional Chinese medicine, the balance of yin and yang foods is believed to determine overall health.[28] In less developed parts of Asia, pregnant women are told to avoid 'hot foods' such as pumpkin and papaya as well as foods that are considered 'too cold' such as cheese, yoghurt and bananas. In rural Ghana, women traditionally abstain from eating these 'hot foods' as well as meat for fear of having deformed babies. Such information is often passed down in families by women who they trust. Many of the dieticians who I spoke to from India and China said that they ate these 'forbidden' foods during pregnancy without any issues at all, despite receiving much criticism and disappointment from their relatives and in-laws.

Scientists are reluctant to start experimenting with pregnant women's diets as it may be considered unethical, but we do tamper with mouse diets. We know that pregnant mice following highly restrictive diets with low fibre alter their gut microbiota, and increase the chances of their offspring developing allergies, obesity and other health problems.[29] Anecdotally, most female dietary or pregnancy experts that I have talked to have had the occasional slice of salami, a little wine or a runny egg while pregnant without stressing. We should be focusing on all the good things to eat, not the rare bad things.

In the Western world, we tend to be overly cautious but we need to draw a line and put risks into perspective. The fear and hype around certain foods and the medical focus on further restricting foods can increase anxiety and interfere with nutrition. For some women, eating anything during pregnancy is a challenge due to severe bouts of morning sickness. Health professionals should focus on improving the diversity, quality and balance of diets and preventing excessive weight gain. Expectant mothers should focus on staying healthy, rather than worrying unduly about taboos around particular foods.

15.
THE ALLERGY EPIDEMIC

Myth: Most of us have a food allergy

Allergies are a modern phenomenon and 21 million Britons and more than 50 million Americans report being affected, many of them food-related.[1] High-profile cases of deaths make the headlines and make us feel more anxious, while schools, restaurants, supermarkets and airlines increase their warnings about allergens. If the trends were to continue, a bag of peanuts could be seen as a weapon of mass destruction. This allergy epidemic has spawned a new and highly lucrative 'free-from' industry, which has seen sales rocket by around 20 per cent per year. But a 2019 study of 8,000 American self-reported food allergy sufferers found only half (10.8 per cent) were truly allergic.[2] Getting an accurate medical diagnosis costs money or takes time, particularly in countries where there is a shortage of specialist allergy doctors.

If you want immediate answers, unconventional internet or high-street tests can be tempting as they provide on-the-spot diagnosis. Browsing the internet, you can now find many ways to solve your problem, by sending back a blood spot, saliva or hair analysis, using the latest technology and science – all available at the click of a button. Saliva tests are even available for your allergic dogs. Other companies have sprung up to offer amazing breakthroughs in this largely unregulated world. Whether you use the internet, a store-based test or a local allergy nutritionist, you can now get a personalised list of foods to avoid and exclusive special free-from products to use.

The problem is that these allergy tests are a useless con, feeding on people's obsessions and anxieties with their diet and health. One journalist tested herself using several different store or online allergy tests and ended up with a long list of 'dangerous' foods, but with no agreement at all between tests.[3] Untrained 'allergists' can often make extra money prescribing 'immune-boosting nutritional supplements', supposedly to enable you to tolerate the allergens. These again are worthless and often expensive, and don't solve the problem if you have a real allergy. You can certainly think yourself into feeling unwell after eating certain foods, and the psychological symptoms are amplified by these tests. So if your allergic friend convinces you that you may also have an allergy to milk, chances are you'll begin to feel unwell after eating it and cut it out of your diet. Essentially, if you believe something will make you ill, it will.

In the medical world, allergies were unheard of up until the 1900s, when a few cases of egg and cow's milk allergy were first documented, and the first formal description was only in 1969. Allergies are on the rise globally, and food allergies are mirroring the dramatic rise in eczema, which is basically a skin allergy.[4] Much of these recent rises are real, but some of it is due to increased publicity and heightened awareness. Confusion between allergies and intolerances and fake tests inflates these figures and leads to misdiagnoses. A food allergy is an abnormal response to a food, triggered by the body's immune system, and symptoms such as wheezing, swelling and vomiting occur within minutes of eating the trigger. Food intolerances are different. They lack a clear definition, and symptoms (abdominal pain, diarrhoea, nausea) can be delayed by as much as forty-eight hours, making them difficult to diagnose. Doctors are becoming better trained on allergies and intolerances, but they don't have the time to delve into causes and symptoms. Inevitably, people look elsewhere to have their diagnoses confirmed, usually by pseudoscience practitioners. Someone with little more than a simplistic weekend training course in acupuncture or kinesiology can set themselves up as an expert and convince a vulnerable person to part with their cash and take potentially harmful treatments or to follow restrictive diets.

A great example is the Vega test. A mixture of acupuncture and homeopathy, it sounds scientific until you realise it measures electrical resistance across the skin while you hold the suspect food in your hand. This is utter nonsense. Another completely useless procedure is hair follicle testing, as hair is not involved at all in immune-based allergic reactions. Other internet companies like Pinnertest and Everlywell offer at-home blood tests for protein antibodies to foods. They test a type of antibody called IgG, which is important in fighting infections and is normally raised in healthy people when you eat. Proper scientific studies have shown that these antibodies are not related in any way to food allergy or intolerance, yet these blood tests show that you are 'allergic' to whatever you eat regularly, even if it is healthy.[5]

You might think that lab tests are well regulated, and that a test that carries for example the European certification (CE) mark, which most do, is clinically robust. All it means is that a test and its packaging conforms to health, safety and environmental standards. It is no guarantee of scientific validity and simply measures what it says it measures. In the US, food allergy tests are marketed as 'laboratory-developed tests', which might seem legitimate and medicalised but you shouldn't be fooled. For as long as they are not making a clinical diagnosis of a disease, they are not regulated. In the UK, the Advertising Standards Authority has forced some allergy-testing companies to remove or modify misleading information. The companies get around this by simply updating their marketing campaigns. The lack of regulation of testing is a growing problem, as with food and vitamin supplements (see chapter 5), and as sales are boosted via the global internet, it cannot be handled by single countries alone. Until rules change, people will continue to be misdiagnosed and misled.

Even in the medical world, the best food allergy tests that we have are IgE blood tests and skin prick tests, and are at best 50 per cent accurate. The IgE blood test measures the levels of a protein that is often raised in some people with allergies. This test is actually clinically meaningful (unlike the tests for the IgG protein). The skin prick test involves pushing tiny quantities of proteins you may be allergic to under the skin with a small needle. If you are allergic, it can trigger a small and controlled allergic reaction in the form of a small red swelling. While these tests are backed by science, they can often be wrong, and we regularly find different results in identical twins with the same allergies. Nearly half of people will test positive to at least one allergen, even if they don't have allergic symptoms, and conversely, many children with symptoms may not have a positive test.[6]

According to my allergy colleagues, skin prick and IgE blood tests are worth doing but only as part of a good medical exam and history by a specialist, who would then check the possible allergy with a supervised food challenge. On their own, they can be dangerous and misleading and a recent study in the US found 80 per cent of people were falsely diagnosed with allergy this way, leading to food avoidance.[7] In contrast, there are no scientifically validated ways of diagnosing food intolerances since, unlike food allergies, they don't involve the immune system. The only way to have some certainty is to conduct a food exclusion diet while keeping a food and symptom diary. Ideally, you should get advice from a dietician and get someone to help you conduct a blinded food re-challenge test with both the real food and dummy foods, so that your anxieties don't bias the result.

Misdiagnosing an allergy or intolerance can be fatal. One teenager told me her GP had diagnosed her for years with a milk allergy based on just IgE blood tests and vague symptoms. For years, she cut out her favourite foods – cheese, yoghurt, cream – from her diet, despite her symptoms worsening. Whenever she returned to her doctor with more symptoms, she was told to eliminate yet more foods from her diet. It wasn't until she switched GPs several years later that her new doctor explained that she had a serious inflammatory bowel disease (Crohn's disease), something that needs proper medical treatment.

Our anxiety about our health and that of our children is fuelled by the fact that extremely rare and fatal allergic reactions are sensationalised, with high-profile cases being reported on the front pages of newspapers. This means that we think severe allergies are much more common than they are in reality. Take the story of Natasha Ednan-Laperouse, for example, a teenager who sadly died as a result of a fatal anaphylactic reaction to sesame seeds while flying from London to Nice in 2016. The culprit? An artichoke, olive and tapenade baguette from Pret a Manger, which contained traces of sesame seeds, something Natasha was highly allergic to and was not clear on the labelling. Some restaurant chains are now so afraid of a diner having an allergic reaction and being sued, that they are advising the one in five people who believe they have food allergies to eat elsewhere.

The chances of dying from anaphylactic reactions are extremely low. In the UK they affect as few as ten people per year. You are far more likely to die from asthma, which kills 1,400 sufferers a year in the UK and around 3,700 in the US. However, statistics like these are not usually reassuring: having a child with a severe food allergy can be extraordinarily stressful for any parent, and allergic episodes are sudden and frightening for everyone involved. It's a highly emotive subject, and this is why allergy doctors stay silent about over-diagnosing allergies. I wrote a piece about allergy in aeroplanes explaining that the small protein component of peanuts that triggers the allergy is not present in the dust released into the air when opening packets. I was overwhelmed by angry, sometimes threatening comments from parents, which made me understand the reticence of my allergy colleagues.

Cutting out entire foods and food groups is becoming a money-making scandal. Clever marketing campaigns, such as the 2019 Oatly advert 'It's like milk, but made for humans', are persuading consumers into following dairy-free diets, with vegan milk sales growing by 30 per cent over the past three years. The cosy relationship between the 'free-from' baby-milk industry and paediatric doctors is making it worse. Between 2006 and 2016, prescriptions of specialist milks for infants with cow's milk allergy increased by 500 per cent, despite no evidence showing any increase in real milk allergy. This over-diagnosis of cow's milk allergy is potentially harmful to mothers and babies, and confuses women on the clear health benefits associated with breastfeeding where allergy is non-existent.

Any form of restrictive diet will put you or your child at high risk of developing nutrient deficiencies or malnutrition, which some national surveys suggest is happening. Some children are having their growth and development stunted by over-restrictive diets, with raw-food and vegan diets even sometimes ending in death due to malnutrition. And there are possible social consequences: a child who is unnecessarily told she can't eat birthday cake, jelly and ice cream may not be invited to as many parties.

If you suspect a food intolerance, by all means experiment with your diet by conducting an exclusion and re-challenge diet, but don't be conned into using these bogus tests. Allergies are more serious (although the chances of having a serious anaphylactic reaction are very low), and medical advice should be sought via your doctor, who will be able to explore your medical history, conduct necessary tests and refer you on to an allergist. Many people don't realise that some allergies such as egg and milk often disappear after a few years, while others such as to peanuts tend to persist. But even these lifelong food allergies may potentially be 'curable' – several colleagues at St Thomas' Hospital have pioneered a new approach where they progressively introduce small amounts of the peanut allergen under the care of an expert allergy doctor. This was successful in a large trial of over 500 people, but so far only in children.[8] This has caused much controversy as to whether or not allergic parents should introduce the peanut allergen early in their child's life.

Probably the worst thing you can do is to be over-cautious and restrict your diet to a few 'safe' foods, as restrictive diets low in diversity and fibre can permanently harm your gut health, especially during pregnancy, potentially worsening your allergies and symptoms.[9] This is a particular problem for children with atopic eczema, where avoidance diets are often harmful.[10] Our obsessions with hygiene, food safety and restrictive diets may have caused many of our current problems, and if we are not careful, our current trends could cause even greater health problems in the future.

16.
THE GLUTEN-FREE
FAD

Myth: Gluten is dangerous

Over the past decade, gluten has acquired a bad reputation. Everyone, regardless of expertise, seems to have an opinion, with celebrities, doctors and nutritionists labelling gluten as unhealthy, unnecessary and potentially dangerous. The wealth of misinformation circulating about gluten and the dearth of good advice on nutrition from professionals has led to the rise in popularity of gluten-free and low-gluten diets. The food industry sees this as a lucrative new market, one currently worth at least $17 billion globally and growing at about 10 per cent a year. With enormous profits at stake, the power of money subtly shapes and influences the debate.

With chicken breasts, shampoo and even water now being labelled as gluten-free to improve sales, the food industry is profiting from consumers' general concerns and misconceptions about gluten. Food companies aren't the only ones – celebrities, health gurus and influencers have bombarded us with their gluten-free offerings, including powerful anecdotes from sports stars such as Novak Djokovic, who, upon achieving the world's number-one tennis slot, credited his success to switching to a gluten-free diet. This seemed to be a powerful testament to forsaking gluten, but shortly afterwards he slipped back down the rankings for several years, showing the dangers of relying on individual anecdotes when it comes to fad diets. There are many similarly dubious stories of celebrities or influencers instantly curing themselves of mysterious illnesses in the same way, when traditional medicine had let them down.

Many of the grains that are typical in most of the world's diets (wheat, rye, barley and oats) contain a protein called gluten. Gluten, which means glue in Latin, is a mix of two smaller storage proteins, gliadin (which gives dough its plasticity) and glutenin (which makes it elastic). Gluten is formed when water is mixed with flour, and it gives dough its characteristic texture, elasticity and shape. These properties can be altered by fermentation, salt, or changing acidity or moisture.

Without a doubt, gluten is a baker's friend, and it is one of the most consumed proteins in the world. It's everywhere. In bread, pasta, pastries, biscuits as well as less obvious sources such as beer, soy sauce and gravy. But for a very small minority (less than 1 per cent) of the population, there is no choice but to avoid all sources of gluten due to a medical diagnosis of coeliac disease or, even more rarely, a wheat allergy. Coeliac disease is a proven allergy to gluten; it is an autoimmune disease where your own immune system attacks your tissues when you eat gluten. Even the smallest morsel of gluten can cause a sufferer to experience a catalogue of debilitating and unpleasant symptoms, including severe diarrhoea, vomiting, severe weight loss, fatigue and anaemia.

For these unlucky individuals, the only way to manage their symptoms is with a gluten-free diet. The damage to the gut is clearly visible under a microscope, which is why suspected patients will have a tiny piece of their small intestine tested as part of the diagnostic process. They will also need specific blood tests, conducted by their doctor, not some miraculous commercially available 'allergy test' purchased on the high street. Crucially, sufferers from coeliac disease need to eat gluten regularly for at least six weeks prior to their medical tests for accurate results.

Coeliac disease is strongly linked to genetic factors, but strangely not everyone with the predisposing gene gets the disease even if they eat gluten. There can be identical twins, both with the coeliac gene and living similar lifestyles, and one can suffer from the disease while the other has zero symptoms. As with so many diseases, our individual patterns of gut microbes could be the key. Coeliac disease is often confused with the much more common IBS (irritable bowel syndrome) or even depression.

True allergy to gluten, then, is a rare phenomenon. So where did all of the fear surrounding it come from? There was an influential study in 2013, conducted on rodents, that showed a correlation between a high-gluten diet and weight gain; there has also been a spate of sensational pseudo-scientific diet books in recent years that lambast gluten as unhealthy, unnatural and bad for our bodies. The negativity surrounding gluten tied in well with emerging trends in the food and health industry and was thus extensively publicised on social media, particularly on 'clean-eating' blogs and similar websites. The rodent study was shared by all and sundry, despite the crucial caveat that the human equivalent of the amount of gluten consumed by the mice would be twenty slices of wholegrain bread per day, an impressive feat for even the most ravenous human. Another study from the same group showed similar results, and the researchers suggested that gluten was somehow adversely affecting the metabolic rate of mice. In a more recent study in 2017, large doses of gliadin protein (the major constituent of gluten) were given to mice, as part of a high-fat diet. While the mice experienced some metabolic and microbial changes, there was no weight gain, and they developed smaller, more efficient fat cells. The lab studies clearly lack consistency, and crucially, all involve genetically inbred lab mice, which rarely translate to human beings.

Despite the current popularity of gluten-free diets (GF), there is no good evidence that avoiding wheat is good for you. Contrary to popular belief, a recent large follow-up study found that long-term dietary intake of gluten is not associated with increased risk of heart disease; conversely, the study found that restriction of gluten led to suboptimal dietary intake of heart-healthy wholegrains, which could increase the risk of heart disease.[1] This study of 100,000 US health professionals, conducted over twenty-six years, found that those with the lowest gluten intakes had a 15 per cent higher chance of having a heart attack. Although observational and so potentially biased, it seems that the extra stress of being on a 'healthy' gluten-free diet may not be good for your heart health after all. Approximately one in ten Brits now say they follow a gluten-free diet, and these figures are even higher in the US, despite less than one in a hundred carrying a medically confirmed diagnosis of coeliac disease. Surveys also show that although most people have heard of gluten-free diets, only 20–50 per cent of people have a clear idea what gluten actually is. Many of these are attempting to follow gluten-free diets, despite lacking this crucial piece of information.

Children with coeliac disease have abnormal gut micro-biomes, with increased levels of the bugs Bacteroides and pro-inflammatory E.Coli, which can be reduced to normal levels by following a strict gluten-free diet. We now know that gut microbes living in our small intestines can produce enzymes that break down gluten into smaller pieces, influencing our variable and individual responses. A small randomised trial of a common probiotic (*Bifido infantis*) in twenty patients suggested coeliac symptoms could be relieved, further pointing to a role of microbes to explain this very curious modern auto-immune disease.

Until recently, coeliac disease was thought to be an exclusively North European illness, although we now know that Americans have the same risk as Europeans (around one in a hundred). Italians – perhaps the biggest consumers of dough in the world – have similar prevalence rates for coeliac disease.[2] Italian coeliacs probably have it tougher than anyone else, with pasta and pizza making up a considerable proportion of the Italian diet. There is some weak evidence that rates of coeliac disease are increasing in some countries, although it's unclear whether this is true or simply a reflection of the 'gluten panic' the food industry is undergoing.

To complicate matters further, people with symptoms triggered by gluten who do not meet the diagnostic criteria for coeliac disease may have a controversial condition – non-coeliac gluten sensitivity (NCGS), a newly established disorder that still lacks a clear clinical definition or diagnostic test.[3] If, after ruling out coeliac disease, you suffer digestive problems and still suspect gluten, you may wish to try a gluten-free diet for six weeks to see if your symptoms disappear and then, crucially, reintroduce gluten into your diet to test whether it is really the culprit. Be warned, though, your chances of being right about gluten are slim. An Italian study in 2015 followed 392 self-reported sufferers of gluten intolerance for two years, asking them to cut out gluten, and then re-introduce it into their diets: 6 per cent had some evidence of coeliac disease; 7 per cent met the criteria for non-coeliac gluten sensitivity; and only one in two hundred had a rare wheat allergy. This left over eight out of ten people who, despite complaining of symptoms after consuming gluten and labelling themselves as gluten intolerant, suffered no obvious adverse effects from the consumption of gluten or wheat. So while gluten sensitivity probably exists in some form, it's far rarer than we're led to believe.

It's common for people to report feeling better on a 'gluten-free' diet, even when, due to a lack of clarity about what gluten is and what products contain it, many people unknowingly consume gluten. The astonishing capacity of our psychological beliefs to have an impact on our physical health, known as the placebo effect, is well documented, and can work in both directions, either improving or worsening symptoms. One in three patients in a clinical trial given a placebo tablet that is reported to often have side effects on the gut, report their gut symptoms get worse; and patients given dummy painkillers report on average a 30 per cent improvement in their pain level. We may be even more suggestible when it comes to food.

Others may feel better on a gluten-free diet because by avoiding gluten, they also eliminate other troublesome foods from their diet, such as beer, wheat and rye, all of which can cause digestive problems in people with IBS. Some people benefit from gluten-free diets as they have to think more carefully about what they are eating, and thus choose healthier foods and avoid random snacking. The benefits will depend on how good or bad your regular starting diet was. This is a similar phenomenon to what many people experience when they try going vegetarian or vegan for the first time. Generally speaking, if you believe a food will make you feel ill or better, the chances are it will, at least in the short term. As people rely more and more on the advice of so-called experts on social media, their diets are becoming increasingly restricted as various different food groups are deemed dangerous or unhealthy based on limited scientific evidence.

While a gluten-free diet may help alleviate symptoms in some people, for others it can lead to nutritional problems. Gluten-free products are typically lacking in vitamin B12, folate, zinc, magnesium, selenium and calcium. Other studies found that gluten-free diets in Spain contained on average more fat and less fibre than comparable diets. It is clear that excluding an entire food group from your diet can reduce fibre and dietary diversity, which also affects our gut microbes, creating the possibility of long-term adverse effects.[4]

Commercially produced gluten-free products are often highly refined and calorie dense due to the complex ingredient substitutions required to get close to the textural properties that gluten provides. A recent study found that gluten-free pasta (which is hard to cook al dente) caused consistently higher blood glucose peaks compared with the original wheat varieties.[5] This may be because a range of highly refined carbohydrate products are used to mimic the texture of wheat, meaning that the sugars are quickly released. The ingredient list on gluten-free foods is often much longer, with many added chemicals that together could be having unknown effects on our body and microbes. Taken together, eating industrial GF foods regularly could in the long term lead to weight gain and a greater risk of diabetes.

Although 65 per cent of Americans surveyed believed GF diets were healthier, there is no good evidence to support this. A 2019 study of 28 healthy volunteers conducted a two-week randomised blinded study on diets with and without gluten. No differences were found between the groups for any symptom.[6] If you are making significant changes to your diet and lifestyle, for example reducing energy-dense, refined foods such as cakes, biscuits and pastries and replacing these with healthier alternatives such as gluten-free grains, fruits and vegetables, some weight loss may occur and you will very likely feel better. On the other hand, if your gluten-free diet consists of highly refined, energy-dense gluten-free products, you may well gain weight and feel worse. These changes in body weight will be unrelated to whether gluten is present in your diet or not. Nutrients aren't the only thing you'll be missing out on – your wallet is likely to be affected too. Gluten-free products are pricey, and you'll need to fork out up to five times more for gluten-free biscuits, breads and pasta compared with standard counterparts.

The large-scale studies suggest that, if anything, eating grains is associated with a lower risk of health problems and obesity. If you need further convincing that it is safe for 99 per cent of us to eat wholegrains, a recent randomised trial in sixty Danish adults for eight weeks found that a diet rich in wholegrains containing gluten reduced both body weight and blood markers of stress (inflammation) in comparison to those on a diet of refined grains.[7] If you experiment and change your diet, be aware that any benefits experienced are unlikely to be due to the gluten itself. Unless you have a medically confirmed diagnosis of coeliac disease or a rare wheat allergy, avoiding gluten is likely 'on average' to do you more harm than good.

17.
ON YOUR BIKE

Myth: Exercise will make you thin

'Keep going!' 'Cycle harder!' 'The more calories you burn while in the orange heart-rate zone, the more fat you will burn and the more you can drink later.' This was the mantra of the exercise class I briefly attended in a London gym a few years ago. The fat-burning myth is one of many that stems from the 1958 idea that a pound of fat is dense and contains 3,500 calories. This extends to the simple idea that if you exercise daily and burn 500 calories per day, you will lose a pound a week and 50 lb in a year. We are told that one of the main reasons we have all got fatter in the last thirty years is that we have become lazy and don't exercise enough. Children no longer walk to school or play regular sports, young people watch too much TV and every day spend hours online at home rather than walking to meet friends, fewer people do any manual work, and many people work at home. The exercise message is aimed at people of all ages from schoolkids to pensioners – go to the gym, walk more, play more sport and expend calories – and your metabolism will improve and the weight will melt away.[1]

To help encourage us we now have wearable devices that tell us when we have reached the magic 10,000 steps and give us permission to have a celebratory snack, recovery sports drink or a beer. My fancy watch bleeps with celebratory tones when I reach the magic number, but this can happen just from me spending a day strolling around large airports like Heathrow, though most of my day was spent sitting in a plane. The 10,000-step goal, which is a nice round number, was invented by a Japanese pedometer company before the 1964 Tokyo Olympics to stop people being lazy and has no scientific basis. The step count doesn't necessarily correlate with increasing your heart rate, and is not affected by intense activities like weight lifting or cycling, so this will miss short bursts of exercise or even just brisk walking, which are likely to be much better for your health. A small study of Scottish postal workers showed that health benefits were only more likely in those walking over 15,000 steps, but there is no clear cut-off.[2]

Many governments continue with the message of exercising to lose weight and even Michelle Obama fronted the 'Let's Move' campaign in the US – but how much is based on science? Does making children play more sport at school prevent them gaining weight and becoming obese adults? Virtually all the data in humans shows little or no difference. A prospective study of 300 school children in Plymouth found no effects of sport on subsequent adolescent weight and a larger study of 6,800 Japanese found no differences in weight in people aged sixty who reported doing lots of sport before the age of thirty.[3] Studies of amateur runners have shown that, despite their best efforts, the average weight slowly increases with age, so they have to run further each year to maintain weight. A follow-up study of fourteen champion TV dieters in the US found exercise was the least successful way to lose weight, and only played a minor role in keeping it off. A number of trials have clearly shown that weight loss is much greater in the dieting groups than exercise groups, and exercise only works to any degree if you eat less at the same time. Studying identical twin pairs who differ in exercise patterns allows a glimpse of what a long-term clinical trial might look like. When we explored our TwinsUK database we found only a 1–2 kg lower weight in the twin who exercised regularly. This underscores the fact that most of our energy expenditure is determined by our genes, and is largely pre-set.

We have always assumed that our ancestors spent their days running around hunting and gathering and this constant activity kept them fit and lean. When I spent a week with the Hadza tribe in Tanzania, the last hunter-gatherers of East Africa, I was surprised. They seemed to be as lazy as us Westerners. They usually have a lie-in, and most of the day they hang around chatting with each other and only go as far as they need to get food, which, as most of the year it is plentiful, is not very far. A team of researchers put activity trackers on them for eleven days as well as measuring their metabolic rates while resting and exercising. They confirmed that they are sedentary for most of the time, walking 4–6 km a day and on average don't expend more calories on physical activity than Westerners. Their resting metabolic rate was also similar to Westerners. The reason they are lean is thanks to their diverse high-fibre, berry and meat diet, not eating much, and avoiding snacks, rather than religiously doing 10,000 steps per day.[4]

There are multiple reasons why exercise doesn't have the dramatic impact we expect. The first is that our expectations are too high. Most of our energy expenditure is predetermined and hard to change. Around 70 per cent is our pre-set resting metabolic rate, which is the energy our cells burn just staying alive; around 10 per cent of our energy is burnt off through the act of digesting food, leaving only about 20 per cent for physical activity, of which around half is for small movements, fidgeting, sitting and standing. This leaves only around 10 per cent of the total expenditure that can be manipulated for most people. This 10 per cent expenditure amenable to change is ten times less than reducing the 100 per cent of energy that goes into your body as food. Even if you manage to force yourself to the gym every day, your body will be fighting your attempts to lose weight. A small amount of fat may be replaced by muscle if you are lucky, which could make you heavier as fat is lighter, but more likely your body will compensate for the potential loss of its energy stores, which it (wrongly nowadays) perceives as dangerous. It does this by making you eat more afterwards and slightly reducing your metabolic rate short term, as well as reducing both your subconscious and conscious activity levels, by making you tired. Rather unfairly, this compensatory mechanism appears greater in overweight people.[5]

To make things worse we routinely over-estimate the amount of energy we burn and underestimate the food we eat afterwards.[6] Even if you don't overeat after exercise (which is unlikely) estimates suggest an average overweight male diligently running for an hour four times a week will only lose a maximum of around 2 kg in a month. Exercise stimulates appetite and just eating an extra slice of pizza can negate forty-five minutes in a lap pool while a Mars bar and an orange juice can wipe out the losses of an exhausting spin class.

Our fixation that exercise can sort out our obesity problems comes from the other major mantra about 'calories in and calories out', which wrongly gave equal weight to both methods of losing weight. Since the 1980s, the food and drink industry has kept this idea going with a sustained subtle advertising campaign influencing the public to think that they are only fat because of slothfulness and if only people did more exercise they could eat and drink as many sugary foods as they wanted. The companies are so rich because the profit margins on sugary snacks and beverages are so huge (around four times that of unprocessed food) that they can afford to spend billions on events like the Olympic Games or the World Cup. As well as conning the public into thinking exercise and sugary beverages go hand in hand, they have spent hundreds of millions funding academics to research the links between physical activity and weight and health. This generous funding has benefited researchers' careers, but deflected them from doing higher-quality research on the far greater dangers of sugar-laden junk foods and snacks. Unfortunately, governments and health departments have been happy to sit back and watch and save their money, as the food industry has set the agenda. When I compared the number of published papers globally over thirty years, there are over twelve times more on physical exercise and weight than on sugar intakes and weight. We also know that industry-funded studies are never entirely unbiased and, despite good intentions of the authors, rarely make conclusions that displease their funders.

Rebranding sugary drinks as sports drinks was a brilliant marketing idea and needed some research to back it up.

Gatorade in the US (now owned by PepsiCo) were the first to do this at scale, and are the equivalent of Lucozade in the UK. Their advertising campaign in the US spread the story that their sugar concoction had at half-time totally transformed the fortunes of the losing American football team in Florida (the Gators), giving them the energy they needed to win the match. They and the other big companies then sponsored research often through intermediaries (such as science institutes or charitable foundations) to show that performances and recovery of athletes were significantly enhanced by their drinks, because of the magic of added minerals and electrolytes. After some investigative journalism, it was revealed in 2015 that Coca-Cola spent over £10 million both funding and influencing research directly in the UK and gave another £5 million via its European Hydration Institute to researchers and influencers like the British Nutrition Foundation, the Obesity Forum, the British Dietetic Association, UKactive and many key advisors of the UK government health and nutrition policy, some of which had publicly expressed doubts over the links between sugar and obesity.[7] A similar story is seen in the USA with millions of dollars spent every year in research funding and lobbying (see chapter 6, page 66).[8]

All the big companies collectively created the myth that constant hydration with their products was crucial to avoid injury and fatigue. They also consistently said that sugar and electrolyte drinks were better than water. Below three hours of exercise, under-hydration has never been shown to be a problem, and as runners are urged to drink as much as tolerable, deaths from over-hydration are now sadly fairly frequent, as opposed to deaths from dehydration, which are non-existent.[9] Most of the previous research conclusions were biased and

any effects trivial, but the idea stuck firmly with the public –
you couldn't do proper exercise without some sugary drink.
Unbiased studies have shown that you don't need any special
drinks or supplements unless you are a professional athlete or
exercising hard for over three hours.[10]

Although there is no evidence that exercise in normal
amounts helps weight loss in most people, there is good evidence it is invaluable for many other common conditions and
should arguably be our number-one prescribed drug. It helps
the efficiency of insulin metabolism, sucking up the sugar into
muscles and so reduces the risk of diabetes. There is also good
evidence that regular bursts of activity, raising the heart rate,
reduces heart disease, high blood pressure and levels of blood
fat. Studies in smaller numbers also show that it can possibly
help depression as much as other therapies, as well as reducing dementia, with a few studies even showing benefit in
schizophrenia.[11] Weight loss is one of the few things exercise
doesn't help with, then, and for that most of us have to eat less
and choose our foods better to match our metabolism and gut
microbes. There are always exceptions to these rules, as we are
all unique genetically and microbially. Our twin studies have
shown that the choice to do exercise is strongly genetic and
naturally some people enjoy it much more than others, who
can find it quite unpleasant.

One recent study looked at 256 students exercising in a gym
and compared their choices of snack before and after. Most
people increased their preference for an unhealthy brownie
compared to an apple after exercise. But one in five people
actually felt less hungry after exercise and were more likely
to change their minds and decline any snack.[12] This means
a few people both enjoy exercise and lack the usual strong

compensatory mechanisms. Don't count on it, though. I still wear my fancy 10,000-step monitor on my wrist, but as several large studies have shown that people wearing them for a year gain more weight than those without, I may remove it when I want to slim down.

18.
FOOD FOR THOUGHT

Myth: Food only impacts the health of
our body, not our mind

'Let food be thy medicine and medicine be thy food.'
Hippocrates knew the importance of food for mood as well as
health, but his message has got lost over the centuries. We have
become obsessed with the power of wonder drugs or mineral
supplements as the one-stop solution to our ills. Melancholia is
the historical term for depression and is most easily described
as losing interest in all the important aspects of life. It is something all of us experience transiently at some time in our lives
for no obvious reason or after some stress, trauma, bereave-
ment or life event, and is quite normal. But some people fail
to rebound from these lows. When this state lasts for more
than a few weeks it is usually termed clinical depression and,
for some, it can go on for years. It can occur at any age and
affects about one in six adults, being increasingly common in
children and rates are much higher in women, even ignoring
the one in seven who experience depression after giving birth.
Depression is common in every country of the world, although

the US is top of the league, with China and Japan being the least depressed. Because of lost work, depression costs us globally over $200 billion a year and is one of the commonest causes of death in young people. Around half of all cases are associated with anxiety, an even commoner problem, which can cause confusion in diagnosis, and gut symptoms are also frequent.

When I worked as a junior doctor in East London on a tuberculosis ward in the 1980s, patients were often confined to the ward for a few months on a cocktail of three antibiotics and were often also depressed. Most were cured, and left very much happier. Some of this happiness, it turned out, was due to a drug called isoniazid which, as well as killing microbes, was shown in subsequent trials to improve mood and lift depression. Because the drug increased levels of the brain chemicals like serotonin and dopamine, it led to the discovery of better, more specialised antidepressant drugs, which have since become the mainstay of treatment. When drugs like Prozac first came out, they were overnight sensations, going on to make the companies billions in profits each year. Companies spent millions on 'gifts and inducements' to doctors to write thousands of prescriptions, even for trivial or short-term problems. Sales continue to rise, and around 15 per cent of UK adults have taken them at some time and 13 per cent of the US population currently take them, over double the rate of cholesterol-reducing drugs like statins. In the UK alone, there are 71 million prescriptions annually and over 300,000 children taking them regularly. Rates of prescriptions have been doubling every ten years in most developed countries as they seem to be treated more like M&Ms rather than hazardous drugs.

The problem is that for many people with genuine depression, these drugs just don't work very well. While for some they can be life-saving, less than half of depressed patients experience any significant improvement, even after increasing doses, and they often have side effects, such as emotional numbness and reduced libido.[1] The many trials performed were funded by the industry, and biased in their conclusions, overestimating the real effects of the drugs in the small numbers they helped. Several large clinical studies have since shown no differences on average between antidepressant drugs compared to behavioural therapy or counselling. This sounds similar to the US opiate scandal, where 75,000 deaths a year occurred after addictive painkillers were prescribed unnecessarily, because of drug industry lobbying and weak medical associations. Although new copy-cat drugs are being developed, there are no new mood-lifting blockbusters in the pipeline and pharma companies have stopped investing. In the race to link specific brain chemicals with specific drugs, we forgot that there may be a bigger picture – the role of gut microbes and the chemicals in food.

Isoniazid is an antibiotic that works against TB and, by altering the gut microbes, may indirectly have a greater effect on mood. The suggestion that our gut might have a role in depression has been ignored until now. A whole range of observational studies have followed hundreds of thousands of people from many countries and consistently found a good diet, one particularly high in plants and seeds and variety, is linked to reduced levels of depression, while diets high in junk food and low fibre and diversity increase the risk.[2] Although these observational studies tried to adjust for many other lifestyle factors that could cause bias, they are unreliable on their own. The good news is that the link between food and mood has been confirmed by recent randomised clinical trials in humans. To the surprise of the investigators, a 2014 study investigating behavioural psycho-therapy and dietary advice in preventing major depression in 247 older people with mild depression found that both methods worked equally well in reducing depressive episodes over two years.[3]

Studies of more severely depressed individuals, randomised to receive either diet support (a Mediterranean-style diet) or social support, showed that improving the diet can significantly improve their mood. As ever, not everyone responded, but the positive changes in the largest study of sixty-seven depressed patients were enough to 'cure' a third of the diet support group after twelve weeks, compared to only 8 per cent of the control group given social support.[4] These impressive results exceed the average response to three months of antidepressant drugs by about threefold. Other short-term diet studies have also generally been positive.[5] Longer-term randomised studies of diet and mood are virtually impossible to perform, but the PREDIMED study of 7,000 overweight Spaniards, which was designed to look at heart disease not mood, although imperfect, is the best we have to date. It found slightly lower depression rates in the group randomised to a high-fat Mediterranean diet with extra vegetables, nuts and olive oil, compared to a conventional lower-fat Western diet over six years, lending further support to the importance of diet.[6]

Until recently, explaining exactly why food or diet changes could possibly affect the brain was tricky, limited to old-fashioned views about specific vitamin or nutrient deficiencies or toxins in food. It needed a new paradigm to help us understand it better, and the discovery of the microbiome fits the bill. It is now clear the complex community of gut microbes that produce thousands of chemicals is key to the link between food and mood. On average, patients with depression have a less diverse set of microbe species, especially in those with the commonest form of depression associated with anxiety. A large recent Flemish-Dutch population study of over 2,000 people showed mood and depression were affected by gut diversity, and the microbes that were missing in depressed people were those producing the key dopamine brain chemicals.[7]

There is a growing consensus that depression is related to elevated levels of inflammation (where our immune system is irritated as if being mildly attacked all the time), and our microbes usually secrete a range of chemicals that keep this inflammation in check and our gut wall healthy. As well as protecting us from inflammation, our microbes also routinely send signals to produce the key brain chemicals like serotonin that lift our mood. This is the same chemical that is increased artificially by modern antidepressants. When we remove gut microbes from sterile mice, the levels of serotonin in their blood and brain plummet and they become depressed (in a mouse way).[8] New studies in mice show that common antidepressants used in humans can reduce levels of key microbes (e.g., Ruminococcus) and this mechanism explains much of their brain action.[9] This could also explain why many people who have the 'wrong' microbes (or diet) may not respond to antidepressants. Evidently, gut microbes are key in influencing our thoughts and emotions and if we can manipulate them safely, there is huge potential for improving our mood and reducing the burden of depression.

It is hard to separate mood from brain function, as our mood depends on our brains producing and recognising the right chemical signals. A common cause of the brain losing its normal function in old age is dementia, where the brain shrinks and memory and emotion are affected. While we don't know the cause of Alzheimer's, the main type of dementia, we are realising it is not just caused by a build-up of plaque in the brain as we thought until recently, but is more a defect of the immune system and exacerbated by a poor diet. Poor diet consistently comes out clearly as the main risk factor following populations who develop dementia. A detailed study of seventy middle-aged people followed with MRI brain scans over three years found that sticking to a Mediterranean diet was significant in predicting losses in brain metabolism.[10] A study of 457 British civil servants over ten years found that those who had the healthiest diets had the least loss in size of key parts of the brain such as the hippocampus, key to emotions and long-term memory.[11] Importantly and more conclusively, randomised trials of people with early memory losses in middle-age given healthy or control diets for three years showed improvements in the hippocampus.[12] Most components of food, except for some polyunsaturated fats, can't cross from the blood into the brain, so much of this effect is likely to be indirect via other chemicals produced in the gut.

As we have discussed before, food is not just carbohydrates, protein and fat, but a mixture of thousands of chemicals that interact with our highly individual gut microbes to modify our brain signals. Schizophrenic patients have severe abnormalities of brain chemicals and abnormal thoughts and delusions. A study in 2019 showed that patients having psychotic thoughts had abnormal gut microbes that when transplanted into lab mice could make them also behave psychotically and altered brain chemicals (such as glutamates and GABA).[13] This suggests the wild idea that schizophrenia could even be partly infectious, and could explain why psychotic patients rarely suffer from viral illnesses and common immune conditions like rheumatoid arthritis.

A similar story is coming from children with autism and milder forms on the autistic/Asperger's spectrum who suffer communication, social and repetitive behaviours. As well as around half of children with spectrum disorder having gut symptoms, small studies from around the world suggest they additionally have abnormal microbes with less diversity compared to control children, with fewer anti-inflammatory microbes linked to other immune disorders in their guts. Some researchers have also hypothesised that herbicides on our food like glyphosate (often known as Roundup, see chapter 21, page 217) have a specific effect on bugs like Clostridia that in some predisposed children lead to autism.[14] Of course, many of the changes in gut bugs may well be related to the idiosyncratic diet habits of spectrum disorder kids.

This, as with depression, can be tricky to unravel, but a small study of eighteen children using a new therapy suggests that microbes could be the cause, not just the consequence. This study, although small and imperfect, has shown improvement in symptoms in a proportion of children over a year given faecal microbial transplants (i.e., poo) from healthy donors or siblings. These can be given by various unsavoury ways: a tube through the nose or up the bottom, or via special dried-poo acid-resistant tablets you swallow (nicknamed crapsules). Another study of twenty-one autism-spectrum kids found that as well as having altered microbes, these species produced altered chemical metabolites in their system, likely to be affecting their behaviour.[15] There is some limited evidence that poo transplants could work to ease depression, from studies both in mice and in a few Japanese patients with irritable bowel syndrome (IBS) whose depressive symptoms improved after transplant.[16]

Larger and better studies are currently ongoing so we may understand the risks and benefits better. But if you have mood or behavioural changes, changing your microbes via your diet should be a priority. Natural probiotics in the form of fermented foods like cheese, yoghurt and kefir (and kimchi and kombucha, if you are adventurous) are likely to be beneficial. As yet, neither yoghurt nor kefir have been specifically studied in mental health, but their microbe ingredients have been tested in randomised trials as part of commercial probiotics used in both mice and humans. Summarising ten small probiotic studies of healthy subjects, an overall improvement in mood and stress was seen in those under the age of sixty-five with less benefit seen in the elderly. In depressed subjects, two out of three studies showed positive benefits over six to eight weeks, with microbes found in the yoghurt predominating.[17] In most of the studies in both mice and humans, the symptoms of anxiety seem to respond best – although we don't know exactly which microbes to recommend. While we still lack good studies, there is reason for optimism that natural yoghurts or probiotic fermented foods in humans could have similar benefits. These encouraging results of so-called 'psychobiotics' are stimulating companies to seek more targeted healthy bacteria that will have the best effects on mood.

Most mental illness develops in some form before the age of fourteen, and so a good varied diet early in life is crucial for prevention. Mothers eating junk food in pregnancy appear to produce children with more behavioural problems, while children who eat poor diets are also at increased risk.[18]

Once a disorder like depression has started, some people will still need traditional antidepressants, but we should all be aware that improving the quality and diversity of diet could play a key role in our mood and preventing dementia.[19] A diverse Mediterranean-style diet with a range of fermented foods to keep your microbes happy is looking like the best present you can offer your brain, both to cheer it up and keep it working well.

19.
THE DIRTY BUSINESS
OF WATER

*Myth: We need to drink eight glasses of water
a day*

We're told we need to drink several litres of water per day to stay hydrated, but more recently we've been told that it's not just the amount that matters, the type of water is important too. And bottled water is big business. After all, it's said to energise you, improve your skin and keep you svelte. The most expensive bottle of mineral water in the world will set you back a mere $60,000 (Acqua di Cristallo), helped by its 24-carat gold bottle and Fijian natural spring water label. If that's a bit pricey, how about a $402 bottle of Kona Nigari water?

We now drink more bottled water than ever before, with its global industry growing rapidly at 10 per cent a year. It's become the most successful bottled beverage in the US, with 50 billion litres sold in 2018, while UK sales have increased fourfold in just twenty years, reaching over 3 billion litres in 2016. By 2025, the global market is estimated to reach $215 billion. People, mainly women, are prepared to pay more for it as they think it is safer, tastes better and is more nutritious.[1] In 2016, water passed an amazing milestone with more bottled water sold than Pepsis, Cokes, Sprites and all other fizzy drinks combined. Should we be celebrating this healthy switch away from sugar or is this a marketing scandal with environmental repercussions?

It's understandable why people worry about the safety of tap water. Globally, waterborne infections, such as cholera, claimed the lives of millions of Westerners until the twentieth century, especially as cities became larger with common water sources. While outbreaks of cholera continue to occur today, they are rare and are confined to developing countries, where sanitation and hygiene levels are generally poor and there is limited access to fresh, clean water.[2] The last reported outbreak of cholera in Western Europe was in 1893 and in the US in 1911, but the fear persists. Several years ago you may well have drunk bottled water while holidaying in countries like Italy, Greece and Spain due to fear of getting ill from their tap water. In the 1970s and early 1980s, investment in infrastructure was lagging, and there were still reported cases of waterborne infections (though not cholera), often in remote islands or villages, which gave tap water a bad reputation. Tourists drank bottled water and, as locals got wealthier, they too switched, believing it to be healthier. Even today, you're given strange looks if you ask for tap water in a restaurant in Greece, Italy or Spain.

More recently, these European countries have received huge amounts of EU funding for water infrastructure, and have some of the world's most advanced water purification and management systems. However, even countries with a history of high-quality water such as the US are not immune to waterborne infections and occasional glitches do occur. In 1993, 400,000 Wisconsin residents became ill with a nasty fungus due to drinking contaminated tap water. Nonetheless, for most people, the chances of getting ill from tap water in developed countries is much less than your chances of dying from a lightning strike or a shark bite. Water companies are required to test regularly their water to ensure that it meets strict standards and must immediately report any problem. Drinks companies have slowly built up unnecessary fears about the safety of tap water to coerce us into buying their expensive products. It still makes sense to do so in some African and Asian countries where water quality isn't as good, but ironically the countries that buy most bottled water have some of the safest, most tested and controlled tap waters on the planet.

Most bottled-water sales are now for still as opposed to carbonated water. There are three main grades of bottled still water; *purified water* is chemically processed tap water, often with a few added minerals. Surprisingly, the source of the purified water and any additives don't have to be declared on the label. Coca-Cola and PepsiCo took ten years to admit that their bestselling Dasani and Aquafina brands were in fact just processed tap water. *Spring water* is derived from underground natural sources and is fit for human consumption straight from the source and cannot be treated with chemicals, with variable mineral composition. Finally, *mineral water* also comes from underground natural sources and is not chemically treated, but must have a minimum mineral or electrolyte content and a constant water flow. Some Italian (San Pellegrino) and French (Badoit) mineral waters have decent amounts of calcium in them (over 180 mg per bottle) and I have recommended them to my vegan osteoporosis patients.

Flavoured mineral waters are another novel category and briefly accounted for a third of UK water sales in 2015, although sales are dropping. They are marketed as healthy alternatives, with artificial fruit flavourings (as opposed to actual fruit) and added sweeteners or sugar that is equivalent to colas. To make them more attractive to consumers, companies add in other recognisable 'healthy' chemical nutrients like aloe vera, turmeric, ginger, omega-3 and vitamin C, allowing them to make health claims that detract from the fact that they don't contain any real fruit. The mix with mineral water is an added allure conferring extra 'magical' and medicinal properties and fantastic profit margins.

It's not just clever marketing campaigns from drink companies that are fuelling sales of mineral water. Some women even bathe in the stuff or spray it on their bodies, believing it to be a cure-all. Mineral water has made its way into expensive beauty products, from powerful anti-wrinkle creams and sunscreens to cleansing body sprays, backed by major PR campaigns, which lack any scientific rigour. The secret springs where the 'miraculous' mineral water is derived from have been around for millennia and spraying mineral water on one's skin (known as balneotherapy) is an ancient Greek practice. Beauty spas and health clinics jumped on the bandwagon a while ago and are selling balneotherapy as an expensive dermatological treatment. The craze for natural untreated products now bizarrely includes buying a bottle of so-called raw water for $36.99 that comes untouched from springs and rivers. While admittedly free from modern chemicals, they are full of microbes, which are billed as probiotic, but it is worth remembering that some water-loving microbes also rarely spread diseases such as cholera.

The bottled-water companies' subtle campaigns to frighten us away from tap water and to have a bottle strapped to our hands 24/7 seems to have worked on our governments as well as ourselves. Bottled-water sales have been greatly aided by government campaigns encouraging people to drink more water. The UK guidelines tell people to drink at least six to eight glasses (1.2 litres) of fluid per day and, in the US and Australia (where temperatures can vary more), at least eight glasses or around 2 litres is recommended. The emphasis is always on getting us to drink up and have more. Is there any evidence behind this recent concern about our going thirsty and drying up? The short answer is no. There is zero support for this.[3] Careful studies looking at intakes over ten years in the elderly failed to show any benefits of extra water on kidney function or mortality either.

One of the reasons we are so afraid of tap water is due to health concerns about chemicals such as chlorine. Chlorine is a natural gas which quickly evaporates from water once added. It's added to tap water to reduce levels of bacteria and reduce infection in many countries, including the US and UK. Levels in your tap water depend on how far your pipes are from the central source where the chlorine was added. Not all countries routinely add chlorine to their water, but this doesn't mean their water is less safe. The Netherlands has three- to fourfold fewer tap-water-related infections than the UK and USA, which are legally required to add chlorine.[4] Different regions add different varieties of chlorine, and some types (e.g. Chloramine) can linger for days. Generally, most tap waters are safe to drink without filtering, but you can buy a water filter to reduce chlorine levels even further. I checked my own taps in North London with some chemical paper tests and found that the water coming from my tap had less than one part per million of free chlorine (i.e., homeopathic doses), making me much more relaxed.

In theory, high levels of chlorine in tap water could be bad for your gut microbes, but negligible amounts reach the microbes – unless you regularly drink from a swimming pool. Chlorine is not the only problem. Unless you buy pricey carbon filters and reverse osmosis machines, your tap water will still contain traces of common pharmaceutical drugs like ibuprofen, oestrogens, antibiotics, and antidepressants.[5] Although levels are low these could have potentially minor cumulative effects: for example, affecting the way our genes function (epigenetics).[6] This might seem like a good reason to switch to bottled water, but a survey in 2013 showed it was no better, and thirteen out of twenty bottle brands also had detectable levels of similar chemicals, including endocrine disruptor chemicals such as bisphenol (BPA).[7] This chemical can have subtle effects on your genes and sex hormones and is now banned in many countries. BPA has been linked to low birth-weight babies and hormone-associated cancers of the breast, prostate and ovary.[8] Manufacturers are responding to public fears by switching to BPA-free plastic, but EU and US regulators say the data is still inconclusive.[9]

We often fixate on chemical 'nasties' to avoid, but one mineral you don't want to miss out on is fluoride. This naturally occurring mineral is found in tap water in varying amounts and is added to some toothpastes. Fluoride has been proven to be effective in reducing tooth decay, and water fluoridation schemes have operated in many countries for over seventy years.[10] The amount of fluoride in your water will vary depending on where you live as some local water supplies naturally contain more fluoride than others. A quarter of UK five-year-olds had tooth decay in 2016–17 and rates continue to rise because of sugar, so it's a no-brainer that governments should be encouraging children and adults alike to drink more fluoridated tap water as it's a safe, easy and effective public health measure and antidote to sugary drinks.

We should all turn our attention to the environmental impact of bottled-water production. Producing bottled water uses 2,000 times more energy than the equivalent volume of tap water. Worse still, it takes around 4 litres of water to purify a single litre of water and over 10 litres to make the plastic to carry it. Add to that the thousands of miles that the bottled water travels to get to cities like London or New York, where the demand is high. Exclusive brands such as Fiji Water claim to be 'carbon neutral' because they re-invest 1 per cent of sales in environmental projects and try and plant trees in Fiji. But these small gestures cannot counteract the enormous energy costs and plastic wastage incurred; the maths doesn't add up.

Many of us who buy bottled water try to recycle the bottles, believing this cuts out most of the environmental waste. What we don't realise is that less than one in five bottles gets recycled globally and even fewer end up as bottles again. In the UK, only 10 per cent of recycled bottles are remade into bottles. While many countries are working hard to reduce their plastic usage, the same cannot be said for China, Indonesia and the Philippines, which have the highest amounts of mismanaged plastic wastage globally.[11] Even the picturesque Cornish beaches in England have been described as 'plastic war zones', and plastic bottles, straws and packaging are regularly washed up on the beaches following storms. The world produces nearly 20,000 bottles per second and they are piling up everywhere.

Most of the plastic ends up floating in oceans like the Pacific. There is one infamous collection known as the Great Pacific Garbage Patch which is twice the size of France. We do get to recycle some of it again – but not as you might imagine. Many of our oceans' fish (including a third of UK fish) are full of plastic microparticles from degraded bottles, which we go on to ingest.[12] These microparticles interact with our bodies and microbes in ways we haven't studied nor can even fully imagine. About 8 million tonnes of plastic is tossed into our oceans each year, most from Asia. Switching to glass bottles would be an obvious solution as glass is easy to recycle, doesn't contaminate the water with chemicals and doesn't enter our food chain. But it is slightly more expensive, which is probably why the multinationals that increasingly control our water supply (Coca-Cola, PepsiCo, Nestlé and Danone) seem reluctant to switch back to glass.

If bottled water is bad for the environment and isn't healthier than tap water, does it at least taste better? Probably not, although it's wholly subjective. Blind tastings have even shown that tap water scores higher than most mineral waters. The wine magazine *Decanter* ran a famous blinded taste comparison with twenty-four bottled waters in London in 2007 using wine-tasting experts. Good old London tap water came in at third, costing less than 0.1p per litre. The losers included New Zealand bottled water, which ranked a dismal eighteenth place despite coming from an extinct volcano and costing 50,000 times more than tap water. Other tastings since have provided similar results with no clear relationship to price, tap water, whether from New York or London, usually scoring rather well. The tests did show that each water tastes different to another, probably due to the varying mineral content. Out of curiosity, I tasted distilled water used in lab experiments and found it strangely bitter and unpleasant. One theory is that we need the natural minerals like salt and calcium in water to match and balance our saliva, and a mismatch causes a taste-bud disconnect.[13]

Getting people to drink more water as opposed to sugary, teeth-rotting drinks such as sodas, fruit juices and cordials can only be a good thing. But expanding the market for plastic bottled tap water using fear and misinformation is good for neither the planet nor our wallets. It's hard to justify the need for bottled water in countries where we're fortunate enough to have some of the most advanced and safe drinking water systems in the world.

Switching to bottled water confers no health benefits; there are far more chemicals and potentially harmful substances in the plastic packaging of bottled water. Before you stick with your expensive bottled waters, do your own blind taste test and see whether you really can taste any difference or, maybe like me, you prefer your tap water. If you do stick with tap, you will both be reducing the global guilt of half a trillion plastic bottles piling up each year on our planet and taking a stand against the power of marketing.

The evidence clearly shows that there is no need to drink more water than your body tells you that you need. Thirst is an extremely well balanced and effective signalling system that we should listen to. Eating fresh fruit and vegetables that are rich in water is a good way to hydrate as part of your meals, and drinking tap water or, if you fancy splashing out, glass-bottled spring water will be absolutely enough. The only exceptions are those who may have a reduced thirst response such as the elderly. If you are fit and healthy, trust your thirst.

20.
JUST A DROP

Myth: Drinking alcohol is always bad for you

Gone are the days when we could enjoy a bottle of wine or a few beers with friends without a vague sense of guilt. The food police have been at us again, with countries like the UK and the Netherlands telling us to drink no more than one glass of wine or pint of beer per day. We're told that drinking any amount of alcohol increases our risk of many diseases, including cancers, liver disease and heart disease. This is in stark contrast to the Mediterranean countries, where moderate alcohol intake is still actively encouraged. There, the pattern of alcohol use is quite different to Anglo-Saxon cultures and you'll see little old ladies sharing a social drink with friends most evenings in the local bars and cafes, and children are encouraged to enjoy a diluted glass of wine with a meal from twelve years old. Booze-loving Britons have reduced their high average alcohol intakes by 20 per cent from 12.6 litres per year in 1990 to 10.4 litres per year in 2017, and a third of UK 16–24-year-olds have begun to shun alcohol altogether.

Similar patterns are emerging throughout hard-drinking Eastern Europe, and within a decade, it's estimated that

Europe will no longer have the highest alcohol intakes, with South Korea and Brazil becoming some of the highest alcohol-consuming countries. Although the average American already drinks less (8.7 litres per year) than the average Brit, they are also abstaining, with beer sales dropping by 1 to 2 per cent per year.[1] Sales of non-alcoholic beer increased globally by a quarter in 2019, and alcohol-free drinks are estimated to be worth over $25 billion annually by 2024. Each week, new alcohol-free bars are opening up in capital cities around the world. But despite this trend, we still have a problem, with over 3 million (one in twenty) deaths globally attributed to alcohol consumption. It has been consistently argued that alcohol, at a population level, is a hundred times more harmful than cannabis, cocaine and heroin.[2]

Drinking heavily is obviously detrimental to your health. In the US, around 10 per cent of alcohol users become addicted, which often leads to liver disease, mental health problems, suicide and early death.[3] Excessive binge drinking costs society a fortune, due to injuries, time off work sick, road accidents, policing, and hospital care from rowdy nights out and alcohol-related diseases. Although we may all be drinking less overall, a recent global survey of over 1,250,000 people in thirty-six countries found that Britons are world-beaters when it comes to binge drinking. On average, those Britons who do drink get inebriated almost once a week. While many drink to excess with friends in social situations for fun, it can lead to antisocial effects including crime, aggression, and physical and sexual violence.

But the story is not so simple. The French drink a lot of alcohol (an average of 11.8 litres per person per year), and yet have the third highest life expectancy compared to other top-earning countries.[4] Multiple observational studies from the late 1970s onwards have consistently suggested that light to moderate drinking (one to two alcoholic drinks per day) is linked to reduced deaths from heart disease compared with being teetotal.[5] This relationship is often described as a J-shaped or U-shaped curve, as both the top and bottom of the charts have increased risk. Although these observational studies are prone to bias, they are the best we have, as randomised controlled trials forcing people to drink or abstain for years are unlikely to happen due to ethical considerations. In the UK, these potentially heart-protective effects were ignored when new data on cancer was published by the Department of Health in 2016.[6] The report said that the risk of getting cancer starts from any level of regular drinking and increases with the amount being drunk. Women were singled out and told there was no safe level, and that drinking just one glass of wine per week would put them at risk of cancer. The guidelines also said that there are no benefits of alcohol for heart health for most people, despite what previous studies have reported. As a result, the UK lowered its alcohol recommendations to one small glass (175 ml) of wine or two small glasses of beer per day for both men and women, equivalent to 14 units (112 g alcohol) maximum per week.

The UK guidelines are now some of the strictest among countries in Europe, and are even lower than the levels set by the drink-conscious USA, where the legal drinking age is twenty-one. In the USA, men are told they can drink two standard alcoholic drinks per day, equivalent to 24.5 units (196 g alcohol) per week, which is nearly double the UK guidelines for men. There is no international consensus as to how much we can drink and what a safe level is. While UK guidelines say there is no safe level of alcohol, in wine-loving Chile, six glasses per day or 49 units per week is considered to be 'low risk'. These discrepancies between countries suggest the scientific conclusions are less robust than we have been told.

According to US research on 333,000 people, those that drank one to two units of alcohol daily lived longer and had 20 per cent less heart disease than teetotallers, and these benefits outweighed any slight increased risk of some cancers.[7] Other studies looked at effects on the brain. One followed 3,000 Americans for thirty years of later life and found that light drinking protected against memory loss and dementia. This was supported by a detailed 2017 study of 550 British civil servants followed for thirty years looking at their brain scans.[8] Their risk was increased if they drank more than six small glasses of wine a week, and mildly reduced if they drank only occasionally. This mild protective effect of light drinking against dementia was confirmed in a larger group of over 9,000 civil servants without scans.[9] Another often overlooked factor is that alcohol brings many people pleasure and can increase social bonding in communities, which plays a key role in increased longevity and better mental health.[10]

Two analyses in 2018 suggested that when all the wider health risks are put together, there was no 'safe limit'. A new analysis of nearly 600,000 drinkers from several countries ran with the headline that, overall, mortality increased steadily with alcohol intake although risk of heart attacks slightly declined. Buried in the depths of the paper was the finding that drinking one to two units per day seemed to be the sweet spot for reducing heart attacks, with 30 per cent lower mortality than for teetotallers.[11] Yet another massive study summarising published data followed a few months later, which was funded by the Gates Foundation, and found that alcohol was linked to twenty-three common adverse health problems, including diseases and road accidents. Again, no safe minimum level for overall mortality was suggested, although it did concede that there were some preventive benefits of moderate drinking on heart disease and diabetes. The analysis deliberately did not show any data on the comparative risk for non-drinkers, as they thought that to do so would be misleading.[12] Yet what is more misleading is to show the data as relative risks, i.e., specific only to groups of drinkers, versus groups of non-drinkers, when absolute risks are more relevant to an individual and account for how rare an outcome is likely to be. If we accept their conclusion that one drink a day increases risk of alcohol-related problems by 0.5 per cent, and then put this into absolute terms, this only means one extra person per 25,000 drinkers would be affected by an event.[13] If they all drank wine, this works out at roughly one extra alcohol-related problem per 1.25 million bottles of wine consumed (based on one bottle a week for 25,000 people for a year). Even I won't get through a million bottles, so I guess my risk from drinking a glass a night is low, and the evidence for total abstinence seems weak.

People break down alcohol differently. If you metabolise the alcohol quickly and efficiently, less gets into the bloodstream, resulting in fewer effects on the body, and you will therefore become less intoxicated. Unfortunately, it's difficult to change how your body breaks down alcohol, as it's mainly determined by factors which you can't control, including ethnicity, age and body size and possibly gender. For example, more than a third of people with East Asian heritage lack a functioning form of the enzyme aldehyde dehydrogenase, which is required for breaking down alcohol.[14] This leads to acetaldehyde building up in the blood, which causes severe, unpleasant facial flushing. In the UK and Australia, men and women are given the same strict alcohol limits, yet in Spain and America, men can drink almost double. Up until the 1980s, most of the studies on alcohol were done on men, partly because alcoholism was considered to be a particularly male problem.[15]

Although we lack large studies, some inconsistent evidence suggests that women are more susceptible to alcohol than men.[16] Women were in 2019 singled out by the media for ignoring the 'fatal link' between booze and cancer.[17] The data suggests that the lifetime risks of breast cancer are increased by 1.5 per cent by drinking two glasses of wine daily (or a double gin and tonic). This means your average risk of 11 per cent may be increased to 12.5 per cent. If you have a strong history of breast cancer, then this small 1.5 per cent difference may influence your decision. But for most women, even if the data on overall harm were precise (which it isn't), the risks of extra alcohol are minimal and are just one of a dozen factors among others, such as weight gain, pregnancies and lack of physical exercise, that go into ascertaining how much risk each woman has of contracting breast cancer. Putting this all together, excess drinking is undoubtedly a problem for both sexes, but there isn't yet enough hard data to conclude that women should drink much less than men.

Up until recently, there's been very little research on the effects of alcohol on the gut microbiome and whether certain alcoholic drinks are better than others. There was a suggestion from a small Spanish trial that red wine was better than gin or water on increasing the gut diversity over a few weeks, and that it also reduced blood pressure.[18] Other studies have shown a major polyphenol in wine, Resveratrol, has its action enhanced by gut microbes. But apart from one other US study showing that alcohol altered microbes in the mouth, there is no long-term population data.[19] Luckily, in our TwinsUK study we have explored the effect of overall alcohol consumption, drinking frequency and type of alcohol consumed (beer and cider, spirits, white wine and red wine) on the gut microbiome in 1,421 UK twins. We then replicated our findings in two other populations from America and Belgium. We found that there was a significant increase in gut diversity in daily red wine drinkers across the three cohorts, while beer and spirits had no effects.[20] White wine showed only a small increase in the same direction, probably because it lacks the high polyphenol content of the grape skin of red wine. Although some artisanal ciders have more polyphenols than red wine, we didn't have enough cider drinkers to be sure of any benefit. So drinking a moderate amount of red wine daily (one to two glasses) is probably good for your gut microbes and could be a significant factor in explaining the health benefits.

Surveys show many people are looking to cut down their alcohol intake, and a recent trend has been to follow a month of abstention such as Dry January, a global charity initiative that started in the UK in 2014 encouraging people to abstain from alcohol for one month, following the excesses of the Christmas season. Over four million Brits and one in five American drinkers intended to abstain in 2019. Follow-up surveys reported 71 per cent of completers said they had more energy and better sleep with no evidence of a rebound in drinking in February and fewer subsequent problems drinking.[21] Experts agree you'll probably be better off health-wise having a day's abstinence each week and drinking a bit less on the days when you do drink, although the UK recommendation to have two to three alcohol-free days per week isn't supported by a lot of evidence. Try it out and see if your concentration improves and your sleep is more restful.

The government like to appear to be helping the nation's health by enforcing restrictions and guidelines on things that in high amounts are clearly harmful, such as alcohol. The guidelines and government actions often contradict each other, similar to the two-faced approach to sugar. With the exception of a few Scandinavian countries, alcohol is cheap and getting cheaper globally. The price has dropped in relative terms in most parts of the world. In the UK, you can buy a standard bottle of vodka in a supermarket for less than £11, or a vodka and fruit mix at £4 per litre. The UK government gets 77 per cent of the cost of a bottle of spirits and it received £11 billion in alcohol tax revenue in 2017, outweighing the estimated health and social costs. In the US, alcohol is even cheaper and a bottle of vodka is under $9, and taxes on booze have reduced across the board by around 30 per cent with a fourfold reduction in cost compared to disposable income since 1980.[22] The tax per unit is less than five cents and it is estimated that each alcoholic drink costs the US taxpayer over $2 for the huge economic costs.[23] Increasing drink prices, especially on the cheapest alcohols, can deter problem drinking and save lives, but as for sugary drinks and processed foods, the global drinks lobby is strong. Governments around the world are hypocritical when it comes to telling people to reduce drinking, while subsidising the costs.

No one is disagreeing that too much alcohol, along with the associated social issues, is bad for us. But some people have been put off drinking a glass of wine a night by inappropriate advice. Remember that all the latest observational studies consistently point towards moderate drinking being beneficial for most people's heart health compared to those who either don't drink, or drink too much. The guidelines around alcohol units or grams of alcohol add to the confusion, especially as the serving size of glasses in many countries, like the US and the UK, have become enormous, where it has doubled in the last thirty years.[24] In contrast, glass sizes have remained much the same in Mediterranean countries. The easiest way to reduce your drinking is to buy smaller glasses for your beer or wine and to have a few alcohol-free days each week. Of course, stopping at one or two (small) glasses isn't always easy, particularly in countries where alcohol is cheap and a major part of the culture. But we should be more honest and transparent about the scale of risk to us as individuals. Clearly governments have a duty protecting the health of the nation, but I think heavy and binge drinking should be targeted, not those relaxing over a leisurely meal with a fine glass of wine. Cheers!

21.
FOOD MILES

Myth: Local food is always best

In the US, the average food product travels 1,500 miles to reach your plate. Britain once produced the most varieties of apples in the world, but the UK now imports 70 per cent, with some travelling over 10,000 miles. In Bangladesh, prawns are one of their biggest foreign exports with 95 per cent transported over 5,000 miles to satisfy our cravings for prawn cocktail, despite us having local equivalents and 40 million Bangladeshis not having enough food to eat. Over 2 billion avocados are being transported similar distances from Mexico, at the cost of local deforestation and over-use of chemicals. Hawaii grows its own sugarcane, but packets of sugar travel 10,000 miles before they arrive back at a Hawaiian cafe. There are huge environmental, social and economic burdens associated with transporting food around the world. These include air pollution, and global warming through increased greenhouse gases (carbon dioxide, methane gas and nitrous oxide), which reduce heat leaving the earth.

Growing concern over these impacts has led to a debate about reducing so-called 'food miles', a term coined in 1992 by Tim Lang that refers to the distance a food travels from producer to ultimate consumer. Conscientious consumers are now buying their food locally in an attempt to counteract some of this damage and help the planet, helping the environment, economy and producers by buying their strawberries or tomatoes from local farm shops or local suppliers, rather than from global supermarket chains. Buying locally means that the food is travelling from the farmer's field straight to your fork, as opposed to being flown several thousand miles across the globe in industrial containers. This sounds simple and can only be a good thing, surely?

Local producers are more likely to try to protect the environment through sustainable farming practices such as limiting usage of pesticides or herbicides (see chapter 21) and supporting the diversity of local wildlife. But the argument that buying locally protects the environment through reducing food miles and carbon emissions is not always true. Contrary to popular belief, buying local strawberries or tomatoes out of season uses similar amounts of energy per kilo compared with importing goods from abroad and isn't any better for the environment. By all means make use of local produce grown in season, so do buy local strawberries during summer months, but don't assume it is better all year round.

Imported fruits and vegetables typically come in huge containers and therefore the sheer volume of food transported counteracts any extra carbon emissions incurred during the longer journey. Estimates suggest sea freight is fifty times more carbon-dioxide efficient than by air. One major downside to relying on mass-produced fruit and vegetables is they are usually produced for volume and consistency, rather than taste, although this doesn't seem to deter most supermarket consumers. In terms of the environmental impact, most imports come via ships, trains and large trucks, all of which usually use less fuel per mile than little local delivery vans. In fact, a UK study comparing farm shops and mass distribution methods found that driving yourself to the local farm shop is worse than having your vegetables delivered in a van to your door in terms of carbon emissions, though cycling is better than both options.[1] Of the estimated 30 billion food miles associated with UK-consumed food, 82 per cent are generated within the country, and over half of these in 2005 were simply due to trips by car from homes to local food shops. So buying your food locally may indeed reduce the distance travelled from field to fork, but any environmental benefits are likely offset by the use of multiple trips of less efficient, smaller, petrol-guzzling vehicles.

The impacts of food transport depend on the mode used. Around a quarter of all HGV transport in the UK is transporting food around the country, and this accounts for about 10 per cent of the CO_2 emissions of all road vehicles. In the UK, 40 per cent of imported fruits and vegetables from Africa are flown by aeroplane. The economy of some of the poorer regions of Africa rely on this UK food trade, so the recent focus on food miles is a concern for their producers. On the surface, it might make sense from an environmental point of view to stop buying fresh produce flown in from Africa, but this would only reduce UK total emissions by less than 0.1 per cent. Despite its high profile, air transport amounts to less than 1 per cent of global food miles. In 2006, two of the big UK retailers introduced a new 'flown' label to deter customers from buying air-flown products and encourage consumption of locally produced foods. This tactic might work better now, but then, it had no apparent impact on sales and it was dropped. It's hard to find any country that tells you the means of transport of the produce on the label. Although advocates of buying locally are correct in saying that producing foods locally reduces the carbon impact of transportation, it's just one part of the much bigger and complex picture and measuring food miles alone is too simplistic.

You might be surprised to learn that buying a locally reared Welsh lamb in the UK is worse for the environment than buying a frozen one. Indeed, lamb which is imported 11,000 miles from New Zealand is, somewhat surprisingly, better for the environment and has a lower carbon footprint.[2] The impact of food transport can be offset, to some extent, if food production is more sustainable than local production methods. New Zealand sheep are generally raised on farms using eco-efficient hydroelectricity (i.e., water power). They also have slightly better weather than the UK, which means that the grass grows for longer and the sheep roam year round needing less feed. For each tonne of lamb produced in the UK, 2,849 kg of CO_2 is emitted compared with only 688 kg in New Zealand-produced lamb. Although there's been some doubt over the accuracy of these figures, most agree that producing lamb in New Zealand is more efficient than in the UK.

Tomatoes are one of the UK's biggest imports, although they are easily grown locally. Importing them from Spain is actually more sustainable in energy-efficiency terms than to produce them in heated greenhouses in the UK. It's also cheaper for the consumer. But even Spain isn't blessed with permanent sunlight, and in our pursuit to have a constant supply of symmetrical tomatoes, Spanish farmers in the southern area of Almeria have steadily expanded their growing season year round for export, outside of their traditional dates (May to October), by using plastic polytunnels. These have proved so financially successful that they now occupy over 64,000 acres of land and the glistening mass is the largest man-made structure visible from space. Using any method in colder climes that seeks to replicate the natural sunlight in other parts of the globe will probably prove counterproductive. Polytunnels, while expensive, are more carbon efficient than greenhouses for, say, growing seasonal fruits such as strawberries; polytunnel-produced strawberries are also more carbon efficient than organically produced varieties using real soil.[3]

If a country is unable to produce a food naturally, should its population continue to eat it at all? More than 80 per cent of the orange juice consumed by Europeans originates from Brazil, the largest orange-juice-producing country. A study several years ago showed that 370,000 acres of land was needed just to grow the oranges necessary to meet the demands of orange-juice consumption in Germany. If everyone in the world drank like the Germans, 32 million acres of land would be needed (equivalent to the size of Greece) just to grow oranges. The same study found that locally produced blackcurrant juice contained similar levels of vitamins as the imported orange juice, but had much lower carbon emissions due to reduced transport. And there are also examples of countries which consume food produced elsewhere, while exporting the same items. The US is one of the largest strawberry producers in the world, yet the majority are exported to Canada and Japan, with Americans relying on cheaper imports from Mexico. Similarly, blueberries are produced seasonally in Connecticut yet the US instead gets cheaper imports from Chile. In these instances, it would be far more sustainable if we lived within our regional food-producing capacities, making use of locally grown berries rather than importing these foods from afar.

Nothing beats the taste of a Mediterranean tomato, grown in a hot, sunny climate, so it would make sense to stick to eating tomatoes imported from Spain during the peak growing season or that you have grown at home, rather than eating local produce from the UK or Netherlands reared in massive artificially lit greenhouses and fed without soil through tubes. We should also eat more seasonal and locally grown plants during the winter months. Happily, growing your own fruit and vegetables is becoming mainstream. There's been a recent rise in urban agriculture, with inner-city schools, community centres and even businesses growing food in city parks, communal parks and on rooftops, as well as in people's gardens, patios and balconies. Allotments are cheaper options for those who are lucky to get them and wish to eat healthily, but struggle to afford fresh, organic local produce. But before we get too excited, we need to be realistic and recognise that we will never be able to feed everyone through urban agriculture and allotments alone; agriculture uses 35–40 per cent of the world's land, yet cities and suburbs occupy just 1 per cent.

Some people have become so fixated on reducing their food miles that they miss the bigger picture. If you want to help the global environment, reduce rising temperatures, and hope to feed the 10 billion people soon to be on our planet, you'd be better off eating more plants, dramatically cutting down on animal products and adopting a flexitarian diet. The effects would be enormous; as we discussed earlier (chapter 9), livestock (mainly beef) contributes around 15 per cent of global greenhouse gas emissions because of the huge amount of land needed to feed them.[4] In the UK, it was estimated that diet change alone would reduce total greenhouse gases by 17 per cent and increase lifespan by an average of eight months.[5] Globally, changing from meat diets to plants would allow us to convert around 76 per cent of our farmland back to the natural environment.[6] For all animal products, later steps in the production chain such as transport, processing and packaging are relatively far less important. For example, transporting the finished milk product generates only 10 per cent of the carbon emissions compared with the milk production stage itself. Growing 1 kg of beef has been estimated to equate in carbon opportunity costs (which includes alternative land use) to one London to New York return flight.[7] Even if these precise figures are disputed (which they are), the relative importance of food sources and diet choices on our planet cannot be ignored.[8]

Food miles is merely one aspect that needs to be considered when assessing the sustainability of a food product. The concept is useful for sparking conversations about carbon emissions and food transportation, but it's time to recognise the bigger picture. The global livestock population is currently growing faster than the human population. By all means, rely on home-grown produce and locally produced, artisan foods as much as you can, but recognise that it's sometimes better for the environment to get your sustainable lamb, tomatoes and bananas from other countries. We should focus on the carbon emissions behind each food item, taking into account modes of transport, production methods, packaging, and the volume of food transported. We also need a global perspective where food production is built around a framework of five factors: climate change, biodiversity loss, land system change, freshwater use and nitrogen and phosphorus flows from fertilisers.

Consumers need accurate and simple information on labels, which will enable them to make informed choices, even if some of the information may not be simple or popular, such as showing a higher carbon footprint of organic-grown tomatoes compared to polytunnel tomatoes. It would be good to see governments and retailers promoting and encouraging people to grow their own foods and eat seasonally, with an emphasis on regional food-producing capacities. In countries like Italy and Spain, consumers accept that most fruit and vegetables come only at certain times of the year, but many other countries have lost this culture.

Eating seasonally requires a change in mindset; you can probably go without fresh strawberries during the winter months, so instead opt for blackberries, sloes or frozen berries. Similarly, you probably don't need a constant supply of avocados, mangos and pineapples. These food needn't be eliminated from your diet, but think of these tropical foods as occasional treats rather than everyday essentials. The same concept applies to meat and fish, most of which are imported from afar and have much bigger impacts. Worrying over where your tomatoes come from is irrelevant if you're still eating meat every day. Think about eating less of it, spending more for better-quality varieties, and making use of regional alternatives. The closer we are to the food, the more likely we are to be able to check its source and quality. Unlike food, drinks other than water are not essential and account for a substantial portion of our carbon footprint, so consider dropping soft drinks. Try and make your own small incremental changes. For example, stop driving to the supermarket, try to reduce your food waste, avoid single-use plastics, buy foods in season and freeze surplus amounts, and grow your own produce where possible. Even if you only manage to do two of these – it will help.

Embrace low-cost vegetables that are maybe less glamorous: root vegetables like parsnips, turnips and swede are grown in abundance in the UK in winter months, and sweet potatoes, sweetcorn, oranges and grapes in the US. Visit your local greengrocer if you need some inspiration or perhaps try a local vegetable-box delivery scheme to be supplied with fresh, seasonal and organic ingredients each week. Without a doubt, supporting good local food producers is important and can be good for the environment as well as our health, increasing the variety of minimally processed foods available and diminishing the power of the supermarkets. But environmental choices need to be put into the bigger picture of saving the planet for the next generation.

22.
SPRAYING THE PLANET

*Myth: Pesticides and herbicides are safe for
our health*

Glyphosate is the world's most popular herbicide, used in over 5 million acres of farmland in the UK and for 90 per cent of US crops. A company named Monsanto developed it originally as a chemical cleaner for tanks and metal pipes, but when it dripped onto the soil, it was found to kill off many common weeds and the company patented it. It was first sold in 1974 and made the company billions of dollars over the next forty years. It was popular not only because it was an effective weed-killer, but because it specifically targeted competing plants, without harming animals. It also meant that the fields didn't need as much ploughing, which helped to reduce soil erosion and carbon dioxide emissions. Importantly, unlike most of its competitors, it was considered safe for humans by authorities across the world and Monsanto boasted it was safer to eat than table salt.

By using this chemical, farmers could improve the efficiency and yields of their crops and keep prices down. Corn seeds were genetically modified so that they were resistant to glyphosate, and could therefore be planted and sprayed at the same time – a win-win for the industry. The chemical is now found in over 750 products and the most well-known form is called Roundup. It's widely used to reduce weeds in fields, gardens and golf courses, and sprayed to dry out crops just before harvesting. It is even being sprayed on some beaches in the North of England to keep weeds down. Most people on the planet are exposed to this chemical at some level, which is unprecedented. This global success has, inevitably, made it a subject of close examination.[1] We are told that it would cost hundreds of millions of dollars and at least ten years to develop a replacement for glyphosate, and if it was phased out, crop yields would drop and carbon dioxide emissions and prices would increase. Many farmers have become totally dependent on herbicides and would be extremely reluctant to change. Monsanto claims that over 800 papers have shown it to be safe, and it has passed reviews from the Environmental Protection Agency (EPA) in the US and its EU equivalents (EFSA), as well as different World Health Organization (WHO) groups.

In 2015, the International Agency for Research into Cancer (IARC), who provide reports for the WHO, shocked the farming world and reversed previous opinions by classifying glyphosate as a probable carcinogen.[2] They reviewed all the available data and found the evidence of cancer in lab animals convincing: seven of the fifteen long-term studies available showed increased risk of tumours, including lymphomas. Although human data at that point was sparse, it too suggested that glyphosate has a cancer-inducing effect. So suddenly, this ubiquitous chemical, a staple in our food chain, was being subject to much greater scrutiny. In the USA, California's environmental agency also classed it as a probable carcinogen in humans, but the Federal Environmental Protection Agency reviewed the evidence and maintained it was still safe. They were under pressure from the powerful crop-spraying lobby, which prevented the more stringent FDA (Food and Drug Administration) committee analysing the evidence in greater depth. In Europe, the EFSA reviewed the evidence and, unlike the IARC report, found no clear evidence of cancer-causing properties and controversially renewed its licence.

IARC and EFSA are still having public spats about their differing opinions, including allegations of conflicts of interest. The IARC lead glyphosate expert failed to declare a $160,000 advisory payment that he had received from a group of European lawyers who he was helping to bring health claims against the manufacturing company.[3] IARC have a reputation for being over-sensitive and as we have discussed previously have said much the same about red meat, bacon, burnt toast and roast coffee beans, even when these scares were not backed up by any human data. We now know that most of the safety data, which comprises millions of pages of reports, comes from the company itself. Recent litigation cases in the US have exposed the weakness of much of these documents, some of which were either simple plagiarism or ghost-written by the company for 'independent' academics. It also turns out the US EPA did in 1985 briefly reclassify glyphosate as a possible carcinogen based on the company's own animal cancer data, but clever lobbying reversed this decision a few years later.

A study of the stored urine levels of 100 elderly subjects in southern California showed levels of glyphosate had increased nearly tenfold in the past thirty years, to levels higher than most Europeans.[4] But even these are supposedly well within safe limits, being apparently thousands of times lower than levels that harm animals. European and US regulators regularly monitor levels of herbicides and pesticides to ensure they are at 'safe' levels, although whether the bar is set too high is an open question. The US 'safe' levels are several times lower than the tougher European ones, but then over fifteen countries worldwide, including several in Europe (such as Germany and Belgium), have unilaterally said they want to ban it outright.[5]

A California jury in 2018 awarded $80 million in damages to a groundsman with an unusual type of blood cancer – non-Hodgkin's lymphoma. He regularly used to spray hundreds of gallons of the weedkiller Roundup containing glyphosate. The jury said there was sufficient evidence that it was a probable causal factor in that case. Recently, a husband and wife who had frequently sprayed their lawns with it and had both developed non-Hodgkin's lymphoma, four years apart, were awarded the not-inconsiderable sum of $2 billion in punitive damages by another California court. Monsanto has been acquired by the chemical giant Bayer for $63 billion, however, so can therefore afford to appeal and fight these lawsuits for several years before any money gets paid. There are another 9,000 cases pending, with Bayer allegedly looking at a $10 billion payout to keep the lawyers happy.

The epidemiological data in humans is neither clear-cut nor consistent, and non-Hodgkin's lymphoma is hard to diagnose and classify precisely, but a 2016 meta-analysis of six studies of variable quality looking at all blood cancers and leukaemias found a slight increased risk of around 30 per cent associated with exposure to glyphosate.[6] More recently, a study in 2019 followed over 300,000 farmers from France, Norway and the USA who were exposed to five to ten times as much as the general public. They found no increased risk for non-Hodgkin's lymphoma overall, but a modest 36 per cent increase in risk for a rare subtype called diffuse large B cell lymphoma.[7] Rates of these blood cancers have not increased noticeably in the last thirty years, so the effects on most people must be subtle, but if you're a keen gardener or farmer ingesting large amounts over many years you may indeed have increased your risk of some cancers. The latest data suggests that other, stronger pesticide chemicals like organophosphates could be much worse than glyphosate and may cause certain forms of immune-related cancers, including non-Hodgkin's lymphoma.[8]

Chemicals like glyphosate are popular because they are supposed to be harmless to humans and other mammals. Glyphosate disrupts a special chemical pathway in plants, which stops them making essential protein building blocks (amino acids), causing them to die. The problem is that microbes that live in soil and human guts have this same chemical pathway. This means microbes and their genes are acutely sensitive to these chemicals, which disrupt their normal metabolism and alter the thousands of chemicals they produce that in turn keep us healthy. Much of the role of the microbiome is to stabilise our immune system and stop it overreacting, so it's not a far stretch to see how, by being exposed over long periods of time, these herbicides and pesticides could alter our immune systems. There is some weak epidemiological evidence to support a slight increase in immune and allergy disorders related to pesticide intakes, particularly in at-risk populations. These include children who may experience developmental problems in early life, and women of childbearing age who are eating non-organic produce.[9] Rodent studies are less convincing, but rat pups fed low doses developed brain and hormone problems, while others show glyphosate alters the gut microbes and leads to anxiety and depression in mice.[10] Recent data shows glyphosate also interferes with the internal microbiomes of honey bees, affecting their health and pollination, and it may be more than coincidence that their numbers are declining rapidly.[11]

The government bodies reassure us that these chemicals are safe for humans at levels used and that the amount in foods are regularly monitored. But levels are rising and the safety thresholds are based on old-fashioned lab animal data, where rodents were given massive doses to see if they developed extra cancers. No one looked for more subtle changes in the human gut microbiome, for example. Should we be worried? Do we need to scrub and peel all our plants, or just switch to more expensive organic fruits and vegetables? Even if you make a conscious effort to reduce your intake of pesticides, low levels are hard to escape. Washing only removes some of the residue and the more the plants (berries in particular) are washed, the less the natural flavour. Peeling is also ineffective for those chemicals that penetrate deeper. Organic products are not completely free of these chemicals either, as they are now in the air, soil and water supply, even if levels are four or five times lower. Many people are rightly sceptical about the certification schemes and value of organic foods. Ironically, one group who do believe in the future of organic produce are the world's biggest food companies. As well as having friendly-looking family farms on their food labels, they are slowly buying up organic producers as they can see the global market taking off, particularly in places like China. With some organic dairy farms in the US now able to house 15,000 cows, this could kick-start an overdue challenge of our outdated and quaint concepts of organic producers.

There's a lack of good studies investigating whether eating organic food regularly is better for you, but a French study followed almost 69,000 people over nearly five years to look at organic foods and cancer risk. It found that regular eaters of sixteen types of organic products were protected against several cancers by about a quarter.[12] Although the study itself was too short and had the usual biases of observational designs, its findings did seem to be more than chance. Again, they found that the risk of non-Hodgkin's lymphoma was reduced in regular organic users, and this was mirrored in a massive population study in the UK of 680,000 women who were followed over nine years for all types of cancer, although the focus wasn't organic foods and the data inferior.[13]

Different countries vary widely in attitudes to chemical-free, organic produce. Over 6 per cent of EU land is now used for organic farming, which contrasts with just 1 per cent of the US. In some countries like Austria, nearly a quarter of produce sold is organic, with Germany as the largest consumer, whereas in the US it is less than 2 per cent. Depending on your tastes in food, you may be exposed to more herbicides than you think. Many common fruits and cereals often have high levels, which are not reduced much by processing, washing or peeling. If you enjoy breakfast but don't like the thought of going organic, you may be interested to know that US and UK government agencies have independently found common breakfast foods to have particularly high levels. Porridge oats (oatmeal) came top of the crop, followed by oat cereals like Cheerios, wholewheat bagels and wholemeal bread, as well as eggs and occasional samples from organic brands.[14] Most healthy bran cereals tested also had moderate or high levels.

We hope to have some good human data from our twins soon, but our early results suggest that people who try and eat healthily by either being vegetarian or by eating extra fresh fruit and vegetables actually have higher levels of pesticides and weedkiller in their blood and urine than those with poorer diets. We all need to think much more about the long-term effects of ingesting unusual chemicals not found in food. We used to not worry too much about pollution or car emissions, but now realise that was folly as these chemicals enter our body via our skin and lungs and affect our brains and many other organs. Some chemicals are clearly best avoided and, although we don't have all the answers yet, think about washing your fruit and vegetables a little bit more, or consider growing your own. And, next time you go shopping, don't necessarily discount that small odd-shaped organic carrot or packet of organic porridge oats, just because they cost a bit more.

23.
DON'T TRUST ME,
I'M A DOCTOR

Myth: Doctors always know best

This book is concerned with thinking about nutrition, diet and food in a different way. It is the antidote to being spoon-fed fairy stories about food that have made us progressively unhealthier and more anxious. It is about realising you are an individual rather than the 'average' person to whom guidelines are aimed. That person does not exist. Of course, the danger of pointing out all these uncertainties or fallacies to you is that you may lose faith in professionals and struggle to know who to trust, particularly when it comes to what to eat. Both traditional and social media still use the status of the doctor to endorse products or advice. Although social-media stars are becoming gurus in their own right, the old-fashioned image of a doctor posing with a stethoscope, white coat and gleaming white teeth often seems to be all you need to sell your message, vitamin supplements or fad diet. Even if the doctor has never practised medicine or seen a single patient in their career.

We trust our doctors for medical advice and the ones we see most frequently are family doctors or GPs, who are usually overworked and stressed with little time to dispense general health or lifestyle advice. Most medical training is directed towards either triaging patients with potentially serious illnesses for referral to specialists or dispensing drugs for common conditions. In most Western countries, only a few days of our six-year medical degree is devoted to nutrition training; 73 per cent of medical schools offer less than the recommended twenty-three hours and most of this is obscure nutrition biochemistry, quickly forgotten. This leaves around two to three hours for anything remotely practical.[1] By the time young doctors do their family practice specialist training, there's no further nutrition training and they've usually forgotten any rare nuggets gleaned at medical school. This sad lack of any practical education or skills in nutrition is true of medical training across the world, according to a recent study, and it is the same in dentistry, nursing and physiotherapy.[2]

Specialist physicians don't fare any better in their nutrition training. Studies of junior doctors training as specialist physicians in the US reported that 75 per cent felt unconfident about discussing basic nutrition with patients, and the figures are believed to be similar in the UK. At my prestigious London university teaching hospital, a bright young doctor in the middle of a five-year training programme in diabetes and endocrinology was sitting in on my osteoporosis clinic. He told me that he and his ten colleagues would only receive sixty minutes' instruction on diets or nutrition over five years, despite the fact that 90 per cent would be caring for type 2 diabetes patients for the rest of their careers, for whom nutritional advice was crucial. Medical training focuses on exciting diagnostic tests and prescription drugs, and yet lifestyle advice, which is often an equally important part of treatment, is completely overlooked. Neither GPs nor hospital doctors are obliged to keep up to date with any changes in diet or nutrition advice as part of their continuing education.

So where does the average doctor get their nutrition advice from? Unlike prescription drugs, there are no friendly pharmaceutical reps to update, educate and provide free marketing materials. No one leaves a basket of nuts and broccoli on the doctor's desk to hand out to patients. The bulk of education comes from the drip-down of government advice and posters showing healthy-food pyramids or 'Eatwell' plates in the waiting rooms. As we have seen, much of this advice is stale, wrong or outdated.

Doctors, just like the general public, are also influenced directly and indirectly by the food industry. For example, Coca-Cola, via the not-for-profit organisation International Life Sciences Institute (ILSI), has embedded itself in the Chinese Health Ministry to ensure that the benefits of exercise are highlighted, rather than diet or ultra-processed food being targeted as a concern. As we saw in chapter 6, Coca-Cola has been exposed in the US for funding research there for the same aim. ILSI, really more a lobby group than anything else, was founded by a former senior vice-president of Coca-Cola and, while it's prone to keeping its operations and funding under wraps, is supported by at least a dozen other giant food companies including PepsiCo, Nestlé and McDonald's, and manages to discreetly and effectively influence public health and doctors across the world.[3]

Few medical doctors want to specialise in nutrition. Lacking in glamour, or clinical role models, it suffers in comparison to those more 'exciting' areas where they can do procedures or prescribe powerful drugs. For those who do realise its importance, there is no infrastructure. Last week I was approached by three young UK doctors all wanting to work on the microbiome and nutrition, but I was unable to help them practically, as no career path exists apart from finding funds for a research PhD. I recently talked to a friend who is a part-time nurse in a large general practice in Devon in the south-west of England with over 25,000 patients. She looks after all the diabetic patients and successfully prescribes low-calorie, low-carbohydrate diets to many, without any support or interest from the thirteen doctors. A study from Newcastle of 300 overweight people with diabetes published in *The Lancet* showed that if patients coped with a low (800) calorie diet for eight weeks, 90 per cent were in remission and could stop their drugs.[4] Even outside trials, some enthusiastic GPs are reporting success in getting 50 per cent of their type 2 diabetes patients off all medications. While not all patients have the willpower or strength to do this, most doctors don't even offer their diabetes patients this new option. They still prefer the easier route at which they are experts, namely giving drugs to slow down the disease and telling patients to avoid fatty foods, until they die prematurely.

The speed at which advice changes is another problem. If food was considered to be a complex chemical drug, it might be taken more seriously. When a meta-analysis showed that a useful anti-inflammatory painkilling drug called Arcoxia increased heart problems by over 30 per cent, it was withdrawn immediately and every doctor informed by letter and email within a week. Whereas in the UK, Public Health England and the National Health Service continue to advise patients to eat a carbohydrate-based breakfast every day to help weight loss, despite clear evidence to the contrary since 2015. I was told back in 2015 that there was a slow process for rewriting guidelines, one that could not be speeded up, and we are still waiting. They claimed many other stakeholders were involved, including the food industry, who obviously may not be keen on people skipping a processed meal and thereby reducing a lucrative source of revenue. The story is the same for vitamin supplements like vitamin D or omega-3 as they get treated like foods not drugs, even though they are proven not to work. Many doctors are allowed to stay out of touch in ways that wouldn't be tolerated for drugs; for instance, many are still concerned about patients eating high-cholesterol foods, such as eggs, even though that myth was dispelled over a decade ago.

The lack of confidence in handling nutritional problems is mirrored in how doctors deal with obese patients. Obesity-related problems are often ignored by doctors for fear of upsetting patients or having complaints made against them. In a recent survey of British family doctors, around one in three believed they had upset patients by discussing their weight and were now frightened to raise the subject. This appears to be a cultural problem, as my French and Belgian colleagues have no qualms about directness with their patients, most of whom don't take offence. Sadly, few family doctors have the confidence, knowledge or time to help patients with diets or food choices, even though over 70 per cent of their patients have lifestyle-related issues. The weight and diets of the health professional may also influence how they behave: if overweight themselves, they are less likely to offer weight-reduction advice, and if they eat poorly, they are less likely to be useful diet educators.[5] While doctors no longer smoke in their surgeries, as they sometimes did when I started, many would be relaxed about having a can of soda or a packet of crisps or biscuits on the desk, and most hospitals still have vending machines full of chocolates and processed snacks.

Efforts are being made to rectify the teaching of medical students in many countries, though changes can be painfully slow due to university politics. Recent surveys suggest the average UK medical student still only gets two to three hours of actual nutrition teaching, but plenty of biochemistry and lectures on drug medications. The average student is likely to know far more about scurvy than obesity, although they will never see a case of the former. I am one of many trying to change this disastrous situation, which even the NHS has realised is an issue but can't change. The other problem is that any improvements we make now to medical training, which some groups are lobbying for,[6] will take up to a decade to filter through to practice. So how do we educate current doctors to give better nutritional and lifestyle advice? Many say they are too busy and stressed to learn any more, but nutrition training should become mandatory as part of their annual continuing education programmes. Many people tell me that their doctors are not always receptive to being given tips by their patients, and others have bought their family doctor a copy of *The Diet Myth* after reading it themselves.

Doctors often don't believe they need to know much about nutrition as they can simply refer patients to nutritionists and dieticians. This is despite US studies showing that poor diet accounted for half of all deaths from heart attacks, strokes and type 2 diabetes.[7] Keeping up to date is another challenge. While some medics are excellent, others are stuck in the past, obediently following advice and guidance that can be over a decade old, preferring checklists to individualisation. Many unfortunately still believe in the religion of the calorie as central to weight control or are wedded to the mysterious world of supplements, which limits their ability to give useful advice on real food.

Another problem in the UK is that dieticians only work in hospitals and treat diseases, nutritionists deal generally with healthy people and nutritional therapists do a mixture of both, but are less regulated. Each country confusingly refers to nutritionists and dieticians differently with no consensus on training. Obesity is not yet classed as a disease, and in many countries dieticians don't see this as a priority. The other obvious problem is the lack of numbers. There are only 8,000 registered dieticians (of which around half are working) in the UK compared to 290,000 doctors. In the US, where a dietician/nutritionist is more flexible in who they treat, there are still only around 90,000 registered compared to over a million doctors – and a hundred million obese customers.

Doctors cannot pretend any longer that it is OK to remain ignorant and simply pass on the health problems that affects one in three of us.

CONCLUSION: HOW TO EAT

It took 2,000 years before Galileo, the father of modern science, reversed the dogma that heavier objects fall to ground faster than lighter objects, when he dropped some himself from the Leaning Tower of Pisa. We are at a similar tipping point in nutrition. All the indications are that enough of us on this crowded planet are receptive to this new view of ourselves and our environment to effect a real change. We need to abandon our misguided reliance on counting calories, following guidelines and believing misleading labels with fat and carbohydrate percentages; we must fight the need to snack or hydrate constantly, and not be frightened of the occasional extended fast or skipped meal. Now we realise that any one of us is very *unlikely* to be average, it becomes obvious that attempting to follow prescriptive and 'average' guidelines, or someone else's special diet, has a high chance of failure. We must think for ourselves and not be conned by marketing into damaging our planet and ourselves by drinking bottled water or using our most valuable land for feeding livestock.

There are some key take-home points from this book: first, we have to be more selective in the information we believe about food, which is mostly fed to us or distorted by people with a vested interest, and often based on weak data or science. We should never believe anyone who says there is only one simple cause or quick fix, which only they know about. We need to ignore or challenge anyone who says cutting out substance X or buying that special supplement will cure us or make us shed pounds. Let us not be deflected from improving our diet by messages that say that we simply need to do our 10,000 steps, or do more walking or yoga, as exercise is a useless tool for weight loss. Researching this book has changed my own views about many foods. It has put me off eating as much fish, most of which are now just farmed animals or endangered species, or drinking bottled water or diet drinks. I am now less concerned about salt and drinking small amounts of wine, but much more concerned about the environmental impacts of what I consume. If I buy food in a packet, I will count the ingredients and judge it accordingly, and try to ignore the misleading low-salt/cal/fat/GF/GM messaging, and instead see it as a signal to avoid the product.

With the many examples in this book, I hope you will be better armed to deal with the hype, scares and misinformation and find what works for you individually. The science of food is definitely not simple, and is becoming increasingly complex, but a good robust message (adapted from Michael Pollan) that should stand the test of time is to *eat a diverse diet, mainly plants, without added chemicals.*

Remember, too, that not all nutritional advice is wrong. There are some areas where nearly all experts agree, before politicians and industry cloud the message. Eating more plants is definitely important as this gives you more fibre, polyphenols and key nutrients. These fruits and vegetables should however replace other food items, not simply be added on top as most guidelines suggest. Not all plants are equal: some have much more polyphenols than others, and as a general rule bright or dark colours are a good sign, including a wide range of berries, beans, artichokes, grapes, prunes, red cabbage, spinach, peppers, chilli, beetroot and mushrooms. Tannins and bitter tastes are another positive sign, in foods such as high-quality coffee, green tea, extra virgin olive oil, dark chocolate and red wine. But we shouldn't just eat kale smoothies every day; the number and variety of different plants is crucial. The more plant species we can eat in a week (ideally twenty to thirty), the healthier and more diverse our gut microbes become, which helps to keep our bodies in good shape. This is not as hard as it sounds and includes eating all parts of the plant – grains, leaves, bulbs, flowers, seeds, nuts, roots, herbs or spices.[1] Keeping these plants as whole as possible, with less mechanical or chemical processing, is also sensible. We can all be more adventurous in our habits, testing our palates on novel or unusual plants and taking advantage of mixed vegetable and meal boxes to experiment with. These have certainly helped me diversify my range of foods and vegetarian cooking techniques.

So-called probiotics or fermented foods are helpful to give our gut community a regular exposure to fresh live microbes. As well as eating good-quality cheese (ideally made with unpasteurised milk), consuming natural full-fat yoghurt regularly is healthy for most people. For an even more concentrated dose of multiple microbes, try fermented milk called kefir or fermented tea (kombucha) or add fermented vegetables like sauerkraut or kimchi. We can make these ourselves at home, which usually have more microbes than commercial versions. Studies show that we need to ingest a small amount of fermented foods regularly (daily or on alternate days) to have a significant effect as the new bugs don't survive in our guts, just the beneficial effect of their chemical signals.

Remember that we need to think of foods in relative not absolute terms. We can choose to increase our plant intake indirectly by reducing something else, like our consumption of meat, fish or potatoes, to allow more room on our plate for other vegetables. There is no doubt that the greatest contribution to saving the planet we can make as individuals is to reduce our meat, fish and dairy consumption. This will leave more land to grow trees or plants to protect and feed us more efficiently. We should all pay more attention to the source, quality and sustainability of the way the animal is raised, which can vary enormously. We should also get to know all the plants we eat much better, for the same reasons. The more plants we eat, the more we should think about the herbicides we are regularly ingesting and consider organic options until we know more about the risks. Where possible we should try buying our plants from a shop we trust, with real people not automated checkouts, where we can ask questions and improve our education.

Wearable devices are set to be part of our lives, recording exercise, sleep, stress, our heart and probably more. By the time you read this, the personal glucose monitors we used in our PREDICT study may be available without a prescription to healthy people, so you could run your own experiments, and the costs may have decreased. The level and precision of the advice provided from personalised food apps produced by commercial companies such as ZOE[2] will continue to improve as more people participate and share their own data. As well as telling you whether you are better off eating porridge or toast for breakfast, or having ice cream, digestive biscuits or chocolate as a treat, eventually you may learn the optimal time to eat your meals, take your exercise and select the foods that boost your energy, increase your metabolic rate and even improve your sleep quality.

If none of these tests or DIY experiments appeal to you, there are still simple ways we can all mitigate our risks. One way is to reduce the number of large metabolic stresses we experience in a day due to increases in our blood fats, insulin and glucose. While a few of us are sensitive to sugar peaks and troughs and can really sense changing energy levels or food cravings, others, like me, find it much harder. Keeping notes or using a food-logging app can help, or just seeing if we can make it to lunch without fainting and needing a KitKat. Although we can't be sure without testing, there are a few rules to follow which will likely reduce our 'average' number of spikes and so reduce the metabolic stress and hunger signals we produce. Reducing the amount of heavily refined carbohydrates we eat is a start, as these, for most of us, have the most easily accessible sugars. Good examples that worked for me were replacing instant porridge oats with steel-cut oats or swapping white bread for sourdough rye bread. Another obvious choice is to avoid sugary drinks and snacks, especially when consumed on their own without food, other than as a special treat. This includes fruit juices and smoothies. Mixing refined carbohydrates with fatty foods like dairy or high-fibre foods also works in many people: e.g., cheese on toast rather than jam, or eating fruit with yoghurt.

Perhaps the most important message that tends to get forgotten is that we should avoid highly or ultra-processed foods as much as possible. Many of the multiple added ingredients are chemicals that alone or in combination are likely to be bad for our long-term health. We know manufacturers manipulate the flavours to make us eat much more than we want to, and many of the key chemicals can interfere with the health of our gut microbes. These include artificial sweeteners, emulsifiers and preservatives. None of these chemicals are naturally present in the food of any of our ancestors and our gut microbes and genes or hormones will not have evolved to deal with them safely. We should also add antibiotics to this list, which are found in tiny amounts in many cheaper forms of meat and some farmed fish.[3] There is increasing evidence that pesticides and herbicides are harmful to our gut microbes too, and this is a good reason to wash your plants and buy organic if you can afford it.

We all get into food ruts, and often repeatedly eat the same breakfast and lunch on work days. Don't just assume that because, like me, you have had the same 'healthy' sandwich for ten years, it must suit you. A good analogy is trying to find the best fuel for your personal engine to make it the most efficient metabolically. Get the fuel right and you can improve your mileage and keep the system clean; get it wrong and your body can run inefficiently and build up unhealthy by-products. But, given that many of us don't know exactly what foods suit us best, it makes sense to vary our diets to lessen the risk of regularly eating chemicals that our body finds it hard to deal with.

It's easy to test yourself by experimenting with your mealtimes. Try, for example, skipping breakfast for a few days and record how you feel with an extended fast, which is increasingly recognised as having health benefits.[4] While writing this book, I spent a day eating three sugary muffins every four hours to test the effects of mealtimes. My sugar graphs swung all over the place and I felt mentally and physically dreadful. Contrary to what textbooks told me that an 'average' person would experience, my biggest and least healthy peak was in the morning, steadily dropping by the evening. This suggested I dealt best with the identical level of carbohydrates later in the day, and my main meal should be the evening. But you may be completely different. You can also try intermittent fasting or exercising at different times of the day, before or after food, or before or after carbohydrates, and see how your body reacts. As individuals, we need to tune in more to our own body's needs, and these will change as we age over time. Life is one big experiment.

Changing our own food habits or those of our family is one thing, changing that of our country or our planet is a different matter. Poor diet is the biggest single factor in causing modern diseases and accounts for nearly half of all deaths,[5] and the billions we pay in health taxes makes it a problem all of us should want to fix. Changing the system, as ever, comes down to politics and cash. The more we eat, the more food companies make money, and the more we eat ultra-processed foods rather than real food, the more money they make. Is it surprising we were told that regular snacking would help us control weight when food companies made fortunes designing novel tasty snacks and also funded the studies to 'prove it'? Is it also surprising that we were told eating a refined carbohydrate breakfast of cereals, oatmeal and orange juice was essential for health and weight control when big food companies were behind the products, studies and advice?

We would not accept it if cigarette or alcohol companies funded most of the studies of their adverse health effects, or influenced future studies, yet we still permit this for food. In

2019, lobbying stopped the USDA diet advisory board reporting in 2020 on scientific evidence on the effects of eating less meat or ultra-processed foods. These recommendations affect directly around a third of the US food supply and influence many other countries.[6] Most countries are less transparent than the USA in these decisions, which take place behind closed doors. We should no longer allow food companies to influence our scientists and advisors through funding, or allow their lobbyists to lean on politicians to block taxes on sodas or junk foods. The costs of poor diet are higher than most realise. In the UK alone this costs 64,000 lives, excluding COVID-19 diet-related deaths, and at least £74 billion to the economy and tax-payer annually, according to the National Food Strategy document.[7] The fact that processed foods are much cheaper than real food is down to our governments. We subsidise unhealthy food through our taxes, and not just from the billions we spend on health. Around a third of all US subsidies go to corn and wheat producers and even food additives key for processed foods are subsidised, whereas most fruit and vegetables receive zero. EU subsidies are similar, with €41 billion in 2018 favouring all the ingredients (including sugar, meat, dairy, soy and animal feed) that are made into unhealthy processed foods which are also bad for our environment. We should lobby to make healthy food cheaper, even if this means taxes on junk food. Many objected to taxes on cigarettes or alcohol when they were first introduced, but with time they were accepted as normal. We can no longer allow politicians to pretend that they are in favour of improving our health by promising cash for new hospitals if they don't openly address the competing issues of continuing subsidies, junk-food taxes, lobbying by industry and the crucial impact on our environment and planet.

We should all be enraged when we hear of multinationals plundering natural rivers and springs in our countries to bottle water in plastic at a thousandfold mark-ups. Especially when these bottles end up floating in our oceans or as microplastics in our fish and ultimately in our guts. Simple taxes of a few pence on plastic bags have made dramatic differences in many countries, and there is no good reason (other than food and beverage company lobbying) this couldn't be extended to other plastic goods and packaging where there are better alternatives.

Food companies have massive marketing budgets, and the promotion of unhealthy food needs to have the same restrictions as smoking or alcohol to protect us. We should follow the lead of countries such as Chile where cartoon animals have been banned from breakfast cereals and other junk foods disguised as healthy. They also added simple black stop labels to allow shoppers to spot ultra-processed foods, instead of the over-complex labels used in most countries that nobody understands. We should also ban the use of false health messages on packets, such as 'with added vitamins' or 'low in fat' unless the companies can actually prove the food is healthier. We desperately need more transparency to better understand what we are eating.

I believe that most government approaches to nutrition are wrong, and competing interests prevent centralised policy changes. No inducements are being created to encourage the consumption of healthier, less processed foods. Governments also have powerful financial pressures facing them. In the UK, the 2018 sugar tax on drinks was a success in many ways, showing that taxation can rapidly change behaviour, and the hope was this would be extended to other high-sugar foods.

Nevertheless, at the same time as the sugar tax, the government passed laws allowing more sugar refining and cheaper imports, driving down the price of sugar for processed foods. Sugar producers also receive a $700 million annual EU subsidy.[8] Unfortunately the 2018 sugar tax may have been a false dawn, as the 2019 government of Boris Johnson succumbed to industry lobbyists and pledged to do a U-turn and end 'sin taxes'. As far as I know, no major government has yet subsidised foods such as vegetables because they are healthy, while their price relative to ultra-processed foods continues to rise globally.

We need more money spent on unbiased food research, to replace the distorting effect of industry funding, and, after a delay of fifty years, it is finally time to target the harmful effects of junk foods with their chemical additives in a rigorous fashion. We still spend far too little on obesity and food research. The US National Institute of Health (like many other countries) spends about ten times more on cancer and three times more on AIDS (HIV) than diabetes and obesity combined, although the latter diseases cost the country far more and affect many times more people.[9] In my view far too much is still spent on small-scale animal research of limited, if any, value, and far too little on large-scale human studies: humans belong to a specific species and we can't improve our nutrition by studying what is the best dog food. If pharma companies can spend billions of dollars to get a single drug to market by showing it works and is not harmful, why can't we allocate similar sums to food or force global food companies, who are now much richer, to do the same?

It is no longer acceptable for health professionals to remain ignorant about nutrition and obesity, even if they learned little in their training. Doctors, nurses and physiotherapists all have an important role to play and, as with smoking, should be the first to visibly change their own habits. In many countries, including the UK and US, nurses and care-workers have high obesity rates[9] and hospitals are full of junk food and vending machines that generate revenue. When did you last visit a dentist with yellow decaying teeth and a waiting room full of sugary treats? We are all paying for these flawed and hypocritical services – and should demand better.

Food is the best medicine, but also the most complex. We can no longer leave something as important as food in the hands of massive corporations, civil servants, bloggers or celebrities. We all need to take personal responsibility to learn more. Education is our main hope. We need to be teaching our children about real and fake food with the same zeal that we teach them how to walk, read and write.

APPENDIX:
TWELVE-POINT PLAN

The whole point of this book is not to tell you how or what to eat, and I have tried my best not to fall into the trap of offering one-size-fits-all guidelines. But if I had to boil down what I've learned into generic advice that could apply to everyone, these statements should be easy to remember and are hard to argue with:

1. Eat diverse foods, mainly plants, without added chemicals
2. Question the science and don't believe quick-fix single solutions
3. Don't be fooled by labels or marketing
4. Understand you are not average when it comes to food
5. Don't get into food ruts: diversify and experiment
6. Experiment with meal timings and skipping meals
7. Use real food, not supplements
8. Avoid ultra-processed foods with over ten ingredients
9. Eat foods to improve gut microbe diversity
10. Reduce regular blood glucose and blood fat spikes
11. Reduce meat and fish consumption and check its sustainability
12. Educate yourself and the next generation in the importance of real food

ACKNOWLEDGEMENTS

Spoon-Fed would not have been possible without the enthusiasm of my agent Sophie Lambert of Conville & Walsh and my amazing editor Bea Hemming of Jonathan Cape, who have both worked closely with me for the last decade. Harriet Smith, who is trained in nutrition, helped enormously as my researcher, pulling together much of the huge amount of papers and data behind my conclusions. A big thanks to Federica Amati and Lucy McCann who helped with all the updates. Many academics and journalists have helped me directly or indirectly, but I would particularly like to thank Tim Lang, Marion Nestle, Bee Wilson, Sarah Berry, Cathy Williamson, Marita Hennessy, Ted Dinan, John Cryan, Adam Fox, Thomas Barber, Caroline Le Roy, Anna Rodriguez, Peter Kindersley, Dariush Mozaffarian, Robin Mesnage, Paul Franks and Kat Arney. Emma Thompson, Greg Wise and Yotam Ottolenghi, Dan Saladino, Nicola Twilley, Cynthia Graber, Zoe Williams as well as all the TwinsUK volunteers – the Mac twins and Turner twins – were great guinea pigs for the study and helped me with useful discussions. I had useful input from John Vincent, Patrick Holden, Helen Browning, Guy Watson, Sebastian Pole, Phil Chowienczyk, Rob Fitzgerald, Lesley Bookbinder, Leora Eisen, and my long-suffering followers on Twitter and Instagram. I also want to thank the many people trying to fight the system and improve nutrition training for medical students and doctors.

I need to thank my own team, especially my loyal assistant Victoria Vazquez, Debbie Hart who effectively runs everything, and the faculty at King's College London for their unwavering support. The co-founders of the company ZOE Global, George Hadjigeorgiou and Jonathan Wolf, were key to this project as were the great team of ZOE Global in London and Boston who also made the amazing ZOE COVID study possible. I was allowed unique access to the latest results of the PREDICT study that was run as an amazing academic–industry collaboration with over seventy people who helped and tolerated my many self-experiments and journey into personalised nutrition. I would also like to thank my PREDICT colleagues Andy Chan at Mass General Hospital and Christopher Gardner at Stanford in the US, as well as Nicola Segata in Italy and José Ordovas at Tufts. The successes of the ZOE COVID app would not have happened without the amazing support of my two King's colleagues, Claire Steves and Seb Ourselin, who shared the risks as well.

My work required the continuing support of King's College London and is mainly funded by the Wellcome Trust, MRC, NHS, CDRF and the Denise Coates Foundation, to whom I'm very grateful.

Lastly, I can't forget my wife and family, as well as close friends, who put up with me, offered advice, and without whom this book would have been finished much sooner.

NOTES

INTRODUCTION

1 Masako, N., 'Dietary walnut supplementation alters mucosal metabolite profiles during DSS-induced colonic ulceration', *Nutrients* (2019); 11(5): 1118

2 J. P. A. Ioannidis, 'The challenge of reforming nutritional epidemiologic research', *JAMA* (2018); 320(10): 969–970

3 D. S. Ludwig, 'Improving the quality of dietary research', *JAMA* (2019)

4 https://blogs.bmj.com/bmj/2019/10/09/bacon-rashers-statistics-and-controversy/

5 Kate Taylor, 'These three companies control everything you buy', *Business Insider* (4 April 2017)

6 Marion Nestle, *Unsavory Truth: How Food Companies Skew the Science of What We Eat,* Basic Books (2018)

7 K. D. Hall, 'Ultra-processed diets cause excess calorie intake and weight gain: an inpatient randomized controlled trial of food intake', *Cell Metabolism* (2019)

8 T. D. Spector, 'Breakfast: a good strategy for weight loss?' *BMJ* (2 February 2019)

9 A. Astrup, 'WHO draft guidelines on dietary saturated and trans fatty acids: time for a new approach?', *BMJ* (2019); 366: l4137

10 A-L. Barabai, 'The Unmapped chemical complexity of our diet', *Nature Food* (2020); 1: 33–37

1: IT'S PERSONAL

1 www.choosemyplate.gov

2 www.nhs.uk/live-well/eat-well/the-eatwell-guide/

3 A. J. Johnson, 'Daily sampling reveals personalized diet-microbiome associations in humans', *Cell Host & Microbe* (2019); 25(6): 789–802

4 joinzoe.com/studies

5 S. E. Berry, 'Decoding human postprandial responses to food and their potential for precision nutrition', PREDICT 1 Study, *Nature Medicine* (2020) (in press)

6 C. M. Astley, 'Genetic evidence that carbohydrate-stimulated insulin secretion leads to obesity', *Clin Chem* (2018): 64(1): 192–200

7 C. D. Gardner, 'Effect of low-fat vs low-carbohydrate diet on 12-month weight loss in overweight adults and the association with genotype pattern or insulin secretion: the DIETFITS randomized clinical trial', *JAMA* (2018) Feb 20; 319(7): 667–679

2: BREAKING THE FAST

1 K. Sievert, 'Effect of breakfast on weight and energy intake: systematic review and meta-analysis of randomised controlled trials', *BMJ* (2019); 364: 142

2 J. A. Betts, 'Is breakfast the most important meal of the day?', *Proceedings of the Nutrition Society* (2016); 75(4): 464–474; and K. Casazza, 'Weighing the evidence of common beliefs in obesity research', *Critical Reviews in Food Science and Nutrition* (2014); 55(14): 2014–2053

3 D. J. Jenkins, 'Nibbling versus gorging: metabolic advantages of increased meal frequency', *New England Journal of Medicine* (1989); 321(14): 929–934

4 https://www.nhs.uk/live-well/eat-well/eight-tips-for-healthy-eating/ (12 April 2019)

5 K. Gabel, 'Effects of 8-hour time restricted feeding on body weight and metabolic disease risk factors in obese adults: a pilot study', *Nutrition and Healthy Aging* (2018); 4(4): 345–353; and R. de Cabo, 'Effects of intermittent fasting on health, aging and disease', *New England Journal of Medicine* (2019); 381: 2541–51

6 K. Casazza, 'Weighing the evidence of common beliefs in obesity research', *Critical Reviews in Food Science and Nutrition* (2014); 55(14): 2014–2053

7 J. Kaczmarek, 'Complex interactions of circadian rhythms, eating behaviors, and the gastrointestinal microbiota and their potential impact on health', *Nutrition Reviews* (2017); 75(9): 673–682

8 K. Adolfus, 'The effects of breakfast and breakfast composition on cognition in children and adolescents: a systematic review', *Advances in Nutrition* (2016); 7(3): 590S–612S

3: CALORIE COUNTING DOESN'T ADD UP

1 J. Levine, 'Energy expenditure of nonexercise activity', *American Journal of Clinical Nutrition* (2000); 72(6): 1451–1454

2 J. A. Novotny, 'Discrepancy between the Atwater factor predicted and empirically measured energy values of almonds in human diet', *Am J Clin Nutr* (2012); 96(2): 296–301

3 R. N. Carmody, 'Cooking shapes the structure and function of the gut microbiome', *Nature Microbiology* (2019); 4(12): 2052–2063

4 https://www.gov.uk/government/statistical-data-sets/family-food-datasets

5 A. Chaix, 'Time-restricted feeding prevents obesity and metabolic syndrome in mice lacking a circadian clock', *Cell Metab* (2019); 29(2): 303–319

6 C. Ebbeling, 'Effects of a low carbohydrate diet on energy expenditure during weight loss maintenance: randomized trial', *BMJ* (2018); 363: k4583

7 C. D. Gardner, 'Effect of low-fat vs low-carbohydrate diet on 12-month weight loss in overweight adults', *JAMA* (2018); 319(7): 667–679

4: THE BIG FAT DEBATE

1 D. Nunan, 'Implausible discussions in saturated fat "research"; definitive solutions won't come from another million editorials (or a million views of one)', *Br J Sports Med* (2019); 53(24): 1512–1513

2 https://www.nhs.uk/live-well/eat-well/the-eatwell-guide/ (28 January 2019)

3 V. W. Zhong, 'Associations of dietary cholesterol or egg consumption with incident cardiovascular disease and mortality', *JAMA* (2019); 321(11): 1081–1095

4 M. Dehghan, 'Associations of fats and carbohydrate intake with cardiovascular disease and mortality in 18 countries from five continents (PURE): a prospective cohort study', *The Lancet* (2017); 390: 2050–2062

5 R. Estruch, 'Primary prevention of cardiovascular disease with a Mediterranean diet supplemented with extra-virgin olive oil or nuts', *New Engl J Med* (2018); 378(25): e34

6 C. N. Serhan, 'Resolvins in inflammation', *J Clin Invest* (2018); 128(7): 2657–2669

7 V. W. Zhong, 'Associations of dietary cholesterol or egg consumption with incident cardiovascular disease and mortality', *JAMA* (2019); 321(11): 1081–1095

8 D. Mozaffarian, 'Dietary and policy priorities for cardiovascular disease, diabetes, and obesity: a comprehensive review', *Circulation* (2016); 133(2): 187–225

9 L. Pimpin, 'Is butter back? A systematic review and meta-analysis of butter consumption and risk of cardiovascular disease, diabetes, and total mortality', *PLOS ONE* (2016); 11(6): e0158118

10 C. D. Gardner, 'Effect of low-fat vs low-carbohydrate diet on 12-month weight loss in overweight adults', *JAMA* (2018); 319(7): 667–679

5: THE SUPPLEMENTS *REALLY* DON'T WORK

1 H. Hemilä, 'Vitamin C for preventing and treating the common cold', *Cochrane Database of Systematic Reviews* (2013) Jan 31; (1): CD000980

2 S. M. Lippman, 'Effect of selenium and vitamin E on risk of prostate cancer and other cancers: the Selenium and Vitamin E Cancer Prevention Trial', *JAMA* (2009); 301(1): 39–51

3 F. Vellekkatt, 'Efficacy of vitamin D supplementation in major depression: a meta-analysis of randomized controlled trials', *Journal of Postgraduate Medicine* (2019); 65(2): 74–80; and D. Feldman, 'The role of vitamin D in reducing cancer risk and progression', *Nature Reviews Cancer* (2014); 14(5): 342–357

4 K. Trajanoska, 'Assessment of the genetic and clinical determinants of fracture risk: genome wide association and mendelian randomisation study', *BMJ* (2018); 362: k3225

5 B. Ozkan, 'Vitamin D intoxication', *Turkish Journal of Pediatrics* (2012); 54(2): 93–98

6 H. A. Bischoff-Ferrari, 'Monthly high-dose vitamin D treatment for the prevention of functional decline: a randomized clinical trial', *JAMA Internal Medicine* (2016); 176(2): 175–183; and H. Smith, 'Effect of annual intramuscular vitamin D on fracture risk in elderly men and women', *Rheumatology* (2007); 46(12): 1852–1857

7 K. Li, 'Associations of dietary calcium intake and calcium supplementation with myocardial infarction and stroke risk and overall cardiovascular mortality in the Heidelberg cohort', *Heart* (2012); 98: 920–925; and J. B. Anderson, 'Calcium intake from diet and supplements and the risk of coronary artery calcification and its progression among older adults: 10-year follow-up of the multi-ethnic study of atherosclerosis (MESA)', *Journal of the American Heart Association* (2016); 5 (10): e003815

8 B. J. Schoenfeld, 'Is there a postworkout anabolic window of opportunity for nutrient consumption?', *Journal of Orthopaedic and Sports Physical Therapy* (2018); 48(12): 911–914

9 M. C. Devries, 'Changes in kidney function do not differ between healthy adults consuming higher- compared with lower- or normal-protein diets: a systematic review and meta-analysis', *Journal of Nutrition* (2018); 148(11): 1760–1775

10 B. M. Burton-Freeman, 'Whole food versus supplement: comparing the clinical evidence of tomato intake and lycopene supplementation on cardiovascular risk factors', *Advances in Nutrition* (2014); 5(5): 457–485

11 S. M. Lippman, 'Effect of selenium and vitamin E on risk of prostate cancer and other cancers: the Selenium and Vitamin E Cancer Prevention Trial', *JAMA* (2009); 310(1): 39–51

12 A. S. Abdelhamid, 'Omega-3 fatty acids for the primary and secondary prevention of cardiovascular disease', *Cochrane Systematic Review* (2018); 7: CD003177

13 J. E. Manson, 'Marine n-3 fatty acids and prevention of cardiovascular disease and cancer', *New England Journal of Medicine* (2019); 380(1): 23–32

14 S.U. Khan, 'Effects of nutritional supplements and dietary interventions on cardiovascular outcomes', *Annals of Internal Medicine* (2019); 171(3): 190–198

6: THE BITTERSWEET HIDDEN AGENDA

1 I. Toews, 'Association between intake of non-sugar sweeteners and health outcomes: systematic review and meta-analyses of randomised and non-randomised controlled trials and observational studies', *BMJ* (2019); 364: k4718

2 E. K. Dunford, 'Non-nutritive sweeteners in the packaged food supply – an assessment across 4 countries', *Nutrients* (2018); 10(2): e257

3 D. G. Aaron, 'Sponsorship of national health organizations by two major soda companies', *American Journal of Preventative Medicine* (2017); 52(1): 20–30

4 J. Gornall, 'Sugar: spinning a web of influence', *BMJ* (2015); 350:h231 infographic https://doi.org/10.1136/bmj.h231

5 M. G. Veldhuizen, 'Integration of sweet taste and metabolism determines carbohydrate reward', *Current Biology* (2017); 27(16): 2476–2485

6 J. E. Blundell, 'Low-calorie sweeteners: more complicated than sweetness without calories', *American Journal of Clinical Nutrition* (2019); 109(5): 1237–1238

7 J. Suez, 'Artificial sweeteners induce glucose intolerance by altering the gut microbiota', *Nature* (2014); 514(7521): 181–186

8 F. J. Ruiz-Ojeda, 'Effects of sweeteners on the gut microbiota: a review of experimental studies and clinical trials', *Advances in Nutrition* (2019); 10: s31–s48

9 K. Daly, 'Bacterial sensing underlies artificial sweetener-induced growth of gut Lactobacillus', *Environmental Microbiology* (2016); 18(7): 2159–2171

10 joinzoe.com

11 K. A. Higgins, 'A randomized controlled trial contrasting the effects of 4 low-calorie sweeteners and sucrose on body weight in adults with overweight or obesity', *American Journal of Clinical Nutrition* (2019); 109(5): 1288–1301

12 K. Olsson, 'Microbial production of next-generation stevia sweeteners', *Microbial Cell Factories* (2016); 15(1): 207

13 joinzoe.com

14 Q. P. Wang, 'Non-nutritive sweeteners possess a bacteriostatic effect and alter gut microbiota in mice,' *PLOS ONE* (2018); 13(7): e0199080

15 M. C. Borges, 'Artificially sweetened beverages and the response to the global obesity crisis', *PLOS Medicine* (2017); 14(1): e1002195

7: NOT ON THE LABEL

1 G. Cowburn, 'Consumer understanding and use of nutrition labelling: a systematic review', *Public Health Nutrition* (2005); 8(1): 21–28

2 C. J. Geiger, 'Health claims: history, current regulatory status, and consumer research', *Journal of the American Dietetic Association* (1998); 98(11): 1312–1314

3 R. DuBroff, 'Fat or fiction: the diet-heart hypothesis', *BMJ Evidence-Based Medicine* (2019); 29 May, p. ii: bmjebm-2019–111180

4 http://www.fao.org/faostat/en/#data/FBS

5 F. Goiana-da-Silva, 'Front-of-pack labelling policies and the need for guidance', *Lancet Public Health* (2019); 4 (1): PE15

6 R. Estruch, 'Primary prevention of cardiovascular disease with a Mediterranean diet', *New England Journal of Medicine* (2013); 368: 1279–1290

7 G. Ares, 'Comparative performance of three interpretative front-of-pack nutrition labelling schemes: insights for policy making', *Food Quality and Preference* (2018); 68: 215–225

8 R. B. Acton, 'Do consumers think front-of-package "high in" warnings are harsh or reduce their control?', *Obesity* (2018); 26(11): 1687–1691

9 M. Cecchini, 'Impact of food labelling systems on food choices and eating behaviours: a systematic review and meta-analysis of randomized studies', *Obes Rev* (Mar 2016); 17(3): 201–10

10 S. N. Bleich, 'Diet-beverage consumption and caloric intake among US adults, overall and by body weight', *American Journal of Public Health* (2014); 104: e72–e78

11 J. Petimar, 'Estimating the effect of calorie menu labeling on calories purchased in a large restaurant franchise in the southern United States: quasi-experimental study', *BMJ* (2019); 367: l5837

12 J. S. Downs, 'Supplementing menu labeling with calorie recommendations to test for facilitation effects', *American Journal of Public Health* (2012); 103: 1604–1609

8: FAST-FOOD PHOBIA

1 C. A. Monteiro, 'NOVA. The star shines bright', *World Nutrition* (2016); 7(1–3): 28–38

2 C. A. Monteiro, 'Household availability of ultra-processed foods and obesity in nineteen European countries', *Public Health Nutrition* (2018); 21(1): 18–26

3 E. M. Steele, 'Ultra-processed foods and added sugars in the US diet: evidence from a nationally representative cross-sectional study', *BMJ Open* (2016); 6: e009892

4 K. Hall, 'Ultra-processed diets cause excess calorie intake and weight gain: an inpatient randomized controlled trial of ad libitum food intake', *Cell Metabolism* (2019); S1550–4131(19): 30248–7

5 J. M. Poti, 'Ultra-processed food intake and obesity: what really matters for health – processing or nutrient content?', *Current Obesity Reports* (2012); 6(4): 420–431

6 L. C. Kong, 'Dietary patterns differently associate with inflammation and gut microbiota in overweight and obese subjects', *PLOS ONE* (2014); 9(10): e109434

7 R. Mendonça, 'Ultraprocessed food consumption and risk of overweight and obesity', *American Journal of Clinical Nutrition* (2016); 104(5): 1433–1440; and D. Mozzaffarian, 'Changes in diet and lifestyle and long-term weight gain in women and men', *New England Journal of Medicine* (2011); 364(25): 2392–2404

8 A. Bouzari, 'Vitamin retention in eight fruits and vegetables: a comparison of refrigerated and frozen storage', *Journal of Agricultural and Food Chemistry* (2015); 63(3): 957–962

9: BRINGING BACK THE BACON

1 http://www.fao.org/faostat/

2 V. Bouvard, 'Carcinogenicity of consumption of red and processed meat', *The Lancet Oncology* (2015); 16(16): 1599–1600

3 'Plant-based meat could create a radically different food chain', *The Economist* (12 October 2019)

4 M. Dehghan, 'Associations of fats and carbohydrate intake with cardiovascular disease and mortality in 18 countries from five continents (PURE): a prospective cohort study', *The Lancet* (2017); 390(10107): 2050–2062

5 X. Wang, 'Red and processed meat consumption and mortality: dose-response meta-analysis of prospective cohort studies', *Public Health Nutrition* (2016); 19(5): 893–905; and A. Etemadi, 'Mortality from different causes associated with meat, heme iron, nitrates, and nitrites in the NIH-AARP Diet and Health Study', *BMJ* (2017); 357: j1957

6 D. Zeraatkar, 'Red and processed meat consumption and risk for all-cause mortality and cardiometabolic outcomes: a systematic review and meta-analysis of cohort studies, *Ann Intern Med* (2019); 171(10): 721–731

7 R. Rubin, 'Blacklash over meat dietary recommendations raises questions about corporate lies to nutrition scientists', *JAMA* (2020)

8 T. D. Spector, 'Bacon rashers, statistics, and controversy', blog.bmj.com (9 October 2019)

9 J. E. Lee, 'Meat intake and cause-specific mortality: a pooled analysis of Asian prospective cohort studies', *American Journal of Clinical Nutrition* (2013); 98(4): 1032–1041

10 E. Lanza, 'The polyp prevention trial continued follow-up study', *Cancer Epidemiology, Biomarkers and Prevention* (2007); 16(9): 1745–1752; and C. A. Thomson, 'Cancer incidence and mortality during the intervention and post intervention periods of the Women's Health Initiative Dietary Modification Trial', *Cancer Epidemiology, Biomarkers and Prevention* (2014); 23(12): 2924–2935

11 V. Bouvard, 'Carcinogenicity of consumption of red and processed meat', *The Lancet Oncology* (2015); 16(16): 1599–1600

12 J. J. Anderson, 'Red and processed meat consumption and breast cancer: UK Biobank cohort study and meta-analysis', *Eur J Cancer* (2018); 90: 73–82

13 D. Średnicka-Tober, 'Composition differences between organic and conventional meat: a systematic literature review and meta-analysis', *Br J Nutr* (2016); 115(6): 994–1011

14 W. Willett, 'Food in the Anthropocene: the EAT-Lancet commission on healthy diets from sustainable food systems', *The Lancet* (2019); 393: 447–92

15 J. Poore, 'Reducing food's environmental impacts through producers and consumers', *Science* (2018); 360(6392): 987–992

16 M. Springmann, 'Options for keeping the food system within environmental limits', *Nature* (2018); 562: 519–525

17 M. Springmann, 'Health-motivated taxes on red and processed meat: a modelling study on optimal tax levels and associated health impacts', *PLOS ONE* (2018); 13(11): e0204139

18 J. L. Capper, 'The environmental impact of beef production in the United States: 1977 compared with 2007', *Journal of Animal Science* (2011); 89: 4249–4261

19 A. Lopez, 'Iron deficiency anaemia', *The Lancet* (2016); 387(10021): 907–16

20 A. Mentre, 'Evolving evidence about diet and health', *The Lancet Public Health* (2018); 3(9): e408–e409; and F. N. Jacka, 'Association of Western and traditional diets with depression and anxiety in women', *American Journal of Psychiatry* (2010); 167(3): 305–311

21 F. N. Jacka, 'Red meat consumption and mood and anxiety disorders', *Psychotherapy and Psychosomatics* (2012); 81(3): 196–198

22 C. A. Daley, 'A review of fatty acid profiles and antioxidant content in grass-fed and grain-fed beef', *Nutrition Journal* (2010); 9(1): 10

23 C. Pelucchi, 'Dietary acrylamide and cancer risk: an updated meta-analysis', *International Journal of Cancer* (2015); 136: 2912–2922

24 J. G. Lee, 'Effects of grilling procedures on levels of polycyclic aromatic hydrocarbons in grilled meats', *Food Chemistry* (2016); 199: 632–638; and A. A. Stec, 'Occupational exposure to polycyclic aromatic hydrocarbons and elevated cancer incidence in firefighters', *Scientific Reports* (2018); 8(1): 2476

25 C. L. Gifford, 'Broad and inconsistent muscle food classification is problematic for dietary guidance in the US', *Nutrients* (2017); 9(9): 1027

26 N. Bergeron, 'Effects of red meat, white meat, and nonmeat protein sources on atherogenic lipoprotein measures in the context of low compared with high saturated fat intake: a randomized controlled trial', *Am J Clin Nutr* (2019) Jun 4: online

27 EFSA, 'Opinion of the scientific panel on food additives, flavourings, processing aids and materials in contact with food (AFC) related to treatment of poultry carcasses with chlorine dioxide, acidified sodium chlorite, trisodium phosphate and peroxy-acids', *European Food Safety Authority* (2006); 4(1): 297

28 Fiona Harvey, 'British supermarket chickens show record levels of antibiotic-resistant superbugs', *The Guardian* (15 January 2018)

29 Felicity Lawrence, 'Revealed: the dirty secret of the UK's poultry industry', *The Guardian* (23 July 2014)

10: FISHY BUSINESS

1 C. A. Raji, 'Regular fish consumption and age-related brain gray matter loss', *American Journal of Preventive Medicine* (2014); 47(4): 444–451

2 M. C. Morris, 'Fish consumption and cognitive decline with age in a large community study', *Archives of Neurology* (2005); 62(12): 1849–1853

3 A. V. Saunders, 'Omega-3 polyunsaturated fatty acids and vegetarian diets', *Medical Journal of Australia* (2013); 1(2): 22–26

4 W. Stonehouse, 'Does consumption of LC omega-3 PUFA enhance cognitive performance in healthy school-aged children and throughout adulthood? Evidence from clinical trials', *Nutrients* (2014); 6(7): 2730–2758; and R. E. Cooper, 'Omega-3 polyunsaturated fatty acid supplementation and cognition: a systematic review & meta-analysis', *Journal of Psychopharmacology* (2015); 29(7): 753–763

5 J. Øyen, 'Fatty fish intake and cognitive function: FINS-KIDS, a randomized controlled trial in preschool children', *BMC Medicine* (2018); 16: 41

6 J. F. Gould, 'Seven-year follow-up of children born to women in a randomized trial of prenatal DHA supplementation', *JAMA* (2017); 317(11): 1173–1175

7 D. Engeset, 'Fish consumption and mortality in the European Prospective Investigation into Cancer and Nutrition cohort', *European Journal of Epidemiology* (2015); 30(1): 57–70

8 L. Schwingshackl, 'Food groups and risk of all-cause mortality: a systematic review and meta-analysis', *American Journal of Clinical Nutrition* (2017); 105(6): 1462–1473

9 M. Song, 'Association of animal and plant protein intake with all-cause and cause-specific mortality', *JAMA Internal Medicine* (2016); 176(10): 1453–1463

10 D. S. Siscovick, 'Omega-3 polyunsaturated fatty acid (fish oil) supplementation and the prevention of clinical cardiovascular disease: a science advisory from the American Heart Association', *Circulation* (2017); 135(15): e867–e884

11 T. Aung, 'Associations of omega-3 fatty acid supplement use with CVD risks: meta-analysis of 10 trials involving 77,917 individuals', *JAMA Cardiology* (2018); 3(3): 225–234

12 A. S. Abdelhamid, 'Omega-3 fatty acids for the primary and secondary prevention of cardiovascular disease', *Cochrane Systematic Review* (2018); 7: CD003177

13 J. E. Manson, 'Marine n–3 fatty acids and prevention of cardiovascular disease and cancer', *New England Journal of Medicine* (2019); 380: 23–32

14 N. K. Senftleber, 'Marine oil supplements for arthritis pain: a systematic review and meta-analysis of randomized trials', *Nutrients* (2017); 9(1): e42

15 A. G. Tacon, 'Global overview on the use of fish meal and fish oil in industrially compounded aquafeeds', *Aquaculture* (2008); 285(1–4): 146–158

16 J. Poore, 'Reducing food's environmental impacts through producers and consumers', *Science* (2018); 360(6392): 987–992

17 Y. Han, 'Fishmeal application induces antibiotic resistance gene propagation in mariculture sediment', *Environmental Science and Technology* (2017); 51(18): 10850–60.

18 Patrick Whittle, 'Plagues of parasitic sea lice depleting world's salmon stocks', *The Independent* (19 September 2017)

19 Shebab Khan, 'Scottish salmon sold by a range of supermarkets in the UK has sea lice up to 20 times the acceptable amount', *The Independent* (29 October 2017)

20 Jen Christensen, ' Fish fraud: what's on the menu often isn't what's on your plate', *CNN* (March 7, 2019)

21 Kimberly Warner, 'Deceptive dishes: seafood swaps found worldwide', *Oceana Report* (7 September 2016)

22 D. A. Willette, 'Using DNA barcoding to track seafood mislabeling in Los Angeles restaurants', *Conservation Biology* (2017); 31(5): 1076–1085

23 Kahmeer Gander, 'Fraudsters are dyeing cheap tuna pink and selling it on as fresh fish in £174m industry', *The Independent* (18 January 2017)

24 R. Kuchta, '*Diphyllobothrium nihonkaiense* tapeworm larvae in salmon from North America', *Emerging Infectious Diseases* (2017); 23(2): 351–353

25 K. Iwata, 'Is the quality of sushi ruined by freezing raw fish and squid? A randomized double-blind trial', *Clinical Infectious Diseases* (2015); 60(9): e43–e48

26 A. Planchart, 'Heavy metal exposure and metabolic syndrome: evidence from human and model system studies', *Current Environmental Health Reports* (2018); 5(1): 110–124

27 E. Oken, 'Fish consumption, methylmercury and child neurodevelopment', *Current Opinion in Pediatrics* (2008); 20(2): 178–183; and S. K. Sagiv, 'Prenatal exposure to mercury and fish consumption during pregnancy and attention-deficit/hyperactivity disorder-related behavior in children', *Archives of Pediatrics and Adolescent Medicine* (2012); 166(12): 1123–1131

28 T. S. Galloway, 'Marine microplastics spell big problems for future generations', *Proceedings of the National Academy of Sciences* (2016); 113(9): 2331–2333

29 A. S. Abdelhamid, 'Omega-3 fatty acids for the primary and secondary prevention of cardiovascular disease', *Cochrane Systematic Review* (2018); 7: CD003177

30 https://friendofthesea.org/; https://fishwise.org/; https://globalfishingwatch.org

11: VEGANMANIA

1 C. Losasso, 'Assessing influence of vegan, vegetarian and omnivore oriented Westernized dietary styles on human gut microbiota', *Frontiers in Microbiol* (2018); 9: 317

2 J. R. Benatar, 'Cardiometabolic risk factors in vegans; A meta-analysis of observational studies', *PLOS ONE* (2018); 13(12): e0209086

3 H. Kahleova, 'Cardio-metabolic benefits of plant-based diets', *Nutrients* (2017); 9(8): 848

4 M. J. Orlich, 'Vegetarian dietary patterns and mortality in Adventist Health Study 2', *JAMA Internal Medicine* (2013); 173(13): 1230–1238

5 V. Fønnebø, 'The healthy Seventh-Day Adventist lifestyle: what is the Norwegian experience?', *American Journal of Clinical Nutrition* (1994); 59(5): 1124S–1129S

6 S. Mihrshahi, 'Vegetarian diet and all-cause mortality: evidence from a large population-based Australian cohort – the 45 and Up Study', *Preventative Medicine* (2017); 97: 1–7

7 P. N. Appleby, 'Mortality in vegetarians and comparable nonvegetarians in the United Kingdom', *American Journal of Clinical Nutrition* (2016); 103(1): 218–230

8 G. Segovia-Siapco, 'Health and sustainability outcomes of vegetarian dietary patterns: a revisit of the EPIC-Oxford and the Adventist Health Study 2 cohorts', *Eur J Clin Nutr* (Jul 2019); 72(Suppl 1): 60–70

9 G. M. Turner-McGrievy, 'A two-year randomized weight loss trial comparing a vegan diet to a more moderate low-fat diet', *Obesity* (2012); 15: 2276–2281

10 E. Fothergill, 'Persistent metabolic adaptation 6 years after "The Biggest Loser" competition', *Obesity* (2016); 24: 1612–1619

11 F. Barthels, 'Orthorexic and restrained eating behaviour in vegans, vegetarians, and individuals on a diet', *Eat Weight Disord* (2018); 23(2): 159–166

12 N. Veronese, 'Dietary fiber and health outcomes: an umbrella review of systematic reviews and meta-analyses', *Am J Clin Nutr* (2018); 107(3): 436–444

13 H. E. Billingsley, 'The antioxidant potential of the Mediterranean diet in patients at high cardiovascular risk: in-depth review of PREDIMED', *Nutrition and Diabetes* (2018); 8(1): 13; and S. Subash, 'Neuroprotective effects of berry fruits on neurodegenerative diseases', *Neural Regeneration Research* (2014); 9(16): 1557–1566

14 M. J. Bolland, 'Calcium intake and risk of fracture: systematic review', *BMJ* (2015); 351: h4580

15 https://waterfootprint.org/en/resources/waterstat/ (November 2019)

16 W. J. Craig and U. Fresán, 'International analysis of the nutritional content and a review of health benefits of non-dairy plant-based beverages', *Nutrients* (2021); 13(3): 842

17 C. Whitton, 'National Diet and Nutrition Survey: UK food consumption and nutrient intakes', *British Journal of Nutrition* (2011); 106(12): 1899–1914

18 P. Clarys, 'Dietary pattern analysis: a comparison between matched vegetarian and omnivorous subjects', *Nutrition Journal* (2013); 12: 82

19 H. Lynch, 'Plant-based diets: considerations for environmental impact, protein quality, and exercise performance', *Nutrients* (2018); 10(12): 1841

20 R. Pawlak, 'The prevalence of cobalamin deficiency among vegetarians assessed by serum vitamin B12: a review', *European Journal of Clinical Nutrition* (2014); 68(5): 541–548

21 L. M. Haider, 'The effect of vegetarian diets on iron status in adults: a systematic review and meta-analysis', *Critical Reviews in Food Science & Nutrition* (2018); 58(8): 1359–1374

22 T. A. Saunders, 'Growth and development of British vegan children', *American Journal of Clinical Nutrition* (1988); 48(3): 822–825; and Mitchell Sunderland, 'Judge convicts parents after baby dies from vegan diet', *Vice* (15 June 2017)

12: MORE THAN A PINCH OF SALT

1 M. Webb, 'Cost effectiveness of a government supported policy strategy to decrease sodium intake: global analysis across 183 nations', *BMJ* (2019); 356: i6699

2 K. Trieu, 'Salt reduction initiatives around the world – a systematic review of progress towards the global target', *PLOS ONE* (2015); 10(7): e0130247

3 'Hidden salt present in popular restaurant meals', *BBC News online* (11 March 2013)

4 A. J. Moran, 'Consumer underestimation of sodium in fast food restaurant meals', *Appetite* (2017); 113: 155–161

5 K. Luft, 'Influence of genetic variance on sodium sensitivity of blood pressure', *Klin Wochenschr* (1987); 65(3): 101–9

6 O. Dong, 'Excessive dietary sodium intake and elevated blood pressure: a review of current prevention and management strategies and the emerging role of pharma-conutrigenetics', *BMJ Nutrition Prevention & Health* (2018); 1: doi: 10.1136

7 N. A. Graudal, 'Effects of low sodium diet versus high sodium diet on blood pressure, renin, aldosterone, catecholamines, cholesterol, and triglyceride', *Cochrane Database Syst Rev* (9 April 2017); 4: CD004022

8 A. J. Adler, 'Reduced dietary salt for the prevention of cardiovascular disease', *Cochrane Database Syst Rev* (2014); 12: CD009217

9 H. Y. Chang, 'Effect of potassium-enriched salt on cardiovascular mortality and medical expenses of elderly men', *Am J Clin Nutr* (2006); 83(6): 1289–96

10 E. I. Ekinci, 'Dietary salt intake and mortality in patients with type 2 diabetes', *Diabetes Care* (2011); 34(3): 703–9

11 R. R. Townsend, 'Salt intake and insulin sensitivity in healthy human volunteers', *Clinical Science* (2007); 113(3): 141–8

12 A. Mente, 'Urinary sodium excretion, blood pressure, cardiovascular disease, and mortality', *The Lancet* (2018); 392(10146): 496–506

13 F. P. Cappuccio, 'Population dietary salt reduction and the risk of cardiovascular disease. A scientific statement from the European Salt Action Network', *Nutr Metab Cardiovasc Dis* (2018); 29(2): 107–114

14 L. Chiavaroli, 'DASH dietary pattern and cardiometabolic outcomes: an umbrella review of systematic reviews and meta-analyses', *Nutrients* (2019); 11(2), pii: E338

15 Caroline Scott-Thomas, 'Salt replacements could be deadly, say renal specialists' *FoodNavigator* (19 March 2009)

16 K. He, 'Consumption of monosodium glutamate in relation to incidence of overweight in Chinese adults: China Health and Nutrition Survey (CHNS)', *Am J Clin Nutr* (2011); 93(6): 1328–36

17 Q. Q. Yang, 'Improved growth performance, food efficiency, and lysine availability in growing rats fed with lysine-biofortified rice', *Sci Rep* (2017); 7(1): 1389

13: COFFEE CAN SAVE YOUR LIFE

1 Boston Collaborative Drug Surveillance Program, 'Coffee drinking and acute myocardial infarction', *The Lancet* (1972); 300(7790): 1278–1281; and H. Jick, 'Coffee and myocardial infarction', *New England Journal of Medicine* (1973); 289(2): 63–67

2 P. Zuchinali, 'Effect of caffeine on ventricular arrhythmia: a systematic review and meta-analysis of experimental and clinical studies', *EP Europace* (2016); 18(2): 257–266

3 M. Ding, 'Long-term coffee consumption and risk of cardiovascular disease: systematic review and a dose-response meta-analysis', *Circulation* (2013); 129(6): 643–659

4 A. Crippa, 'Coffee consumption and mortality from all causes, CVD, and cancer: a dose-response meta-analysis', *Am Journal of Epidemiology* (2014); 180(8): 763–775

5 J. K. Parker, 'Kinetic model for the formation of acrylamide during the finish-frying of commercial French Fries', *J. Agricultural and Food Chemistry* (2012); 60(32): 9321–9331

6 Hannah Devlin, 'How burnt toast and roast potatoes became linked to cancer', *The Guardian* (27 January 2017)

7 B. Marx, 'Mécanismes de l'effet diurétique de la caffeine', *Médecine Sciences* (2016); 32(5): 485–490

8 Q. P. Liu, 'Habitual coffee consumption and risk of cognitive decline/dementia: a systematic review and meta-analysis', *Nutrition* (2016); 32(6): 628–636; and G. W. Ross, 'Association of coffee and caffeine intake with the risk of Parkinson disease', *JAMA* (2000); 283(20): 2674–2679

9 C. Pickering, 'Caffeine and exercise: what next?', *Sports Medicine* (2019); 49(7): 1007–1030

10 J. Snel, 'Effects of caffeine on sleep and cognition', *Progress in Brain Research* (2011); 190: 105–117

11 A. P. Winston, 'Neuropsychiatric effects of caffeine', *Advances in Psychiatric Treatment* (2005); 11(6): 432–439

12 M. Lucas, 'Coffee, caffeine, and risk of depression among women', *Archives of Internal Medicine* (2011); 171(17): 1571–1578

13 M. Lucas, 'Coffee, caffeine, and risk of completed suicide: results from three prospective cohorts of American adults', *World Journal of Biological Psychiatry* (2012); 15(5): 377–386

14 C. Coelho, 'Nature of phenolic compounds in coffee melanoidins', *Journal of Agricultural and Food Chemistry* (2014); 62(31): 7843–7853

15 D. Gniechwitz, 'Dietary fiber from coffee beverage: degradation by human fecal microbiota', *Journal of Agricultural and Food Chemistry* (2007); 55(17): 6989–6996

16 M. A. Flaten, 'Expectations and placebo responses to caffeine-associated stimuli', *Psychopharmacology* (2003); 169(2): 198–204; and C. Benke, 'Effects of anxiety sensitivity and expectations on the startle eyeblink response during caffeine challenge', *Psychopharmacology* (2015); 232(18): 3403–3416

17 L. Mills, 'Placebo caffeine reduces withdrawal in abstinent coffee drinkers', *Psychopharmacology* (2016); 30(4): 388–394

18 EFSA, 'EFSA opinion on the safety of caffeine' (23 June 2015)

19 B. Teucher, 'Dietary patterns and heritability of food choice in a UK female twin cohort', *Twin Research and Human Genetics* (2007); 10(5): 734–748

20 A. G. Dulloo, 'Normal caffeine consumption: influence on thermogenesis and daily energy expenditure in lean and postobese human volunteers', *American Journal of Clinical Nutrition* (1989); 49(1): 44–50

21 M. Doherty, 'Effects of caffeine ingestion on rating of perceived exertion during and after exercise: a meta- analysis', *Medicine and Science in Sports* (2005); 15(2): 69–78

14: EATING FOR TWO

1 https://www.nhs.uk/conditions/pregnancy-and-baby/foods-to-avoid-pregnant/ (23 January 2017); and https://www.acog.org/Patients/FAQs/Nutrition-During-Pregnancy? (February 2018)

2 J. Rhee, 'Maternal caffeine consumption during pregnancy and risk of low birth weight: A Dose-response meta-analysis', *PLOS ONE* (2015); 10(7): e0132334

3 L. Holst, 'Raspberry leaf – should it be recommended to pregnant women?', *Complementary Therapies in Clinical Practice* (2009); 15(4): 204–208

4 D. A. Kennedy, 'Safety classification of herbal medicines used in pregnancy in a multinational study', *BMC Complementary Alternative Medicine* (2016); 16: 102

5 E. P. Riley, 'Fetal alcohol spectrum disorders: an overview', *Neuropsychology Review* (2013); 21(2): 73–80

6 U. S. Kesmodel, 'The effect of different alcohol drinking patterns in early to mid pregnancy on the child's intelligence, attention, and executive function', *BJOG* (2012); 119(10): 1180–1190

7 S. Popova, 'Estimation of national, regional, and global prevalence of alcohol use during pregnancy and fetal alcohol syndrome: a systematic review and meta-analysis', *The Lancet* (2017); 5: e290–e299

8 R. F. Goldstein, 'Association of gestational weight gain with maternal and infant outcomes: a systematic review and meta-analysis', *JAMA* (2017); 317(21): 2207–2225

9 https://www.nice.org.uk/guidance/ph27/chapter/1-Recommendations#recommendation-2-pregnant-women (July 2010)

10 C. H. Tam, 'The impact of maternal gestational weight gain on cardiometabolic risk factors in children,' *Diabetologia* (2018); 61(12): 2539–2548

11 V. Allen-Walker, 'Routine weighing of women during pregnancy — is it time to change current practice?', *BJOG* (2015); 123(6): 871–874

12 F. Hytten, 'Is it important or even useful to measure weight gain in pregnancy?' *Midwifery* (1990); 6(1): 28–32; and M. G. Dawes, 'Repeated measurement of maternal weight during pregnancy. Is this a useful practice?', *BJOG* (1991); 98(2): 189–194

13 https://www.nhs.uk/common-health-questions/pregnancy/how-much-weight-will-i-put-on-during-my-pregnancy/ (18 October 2018)

14 K. V. Dalrymple, 'Lifestyle interventions in overweight and obese pregnant or postpartum women for weight management: a systematic review', *Nutrients* (2018); 10(11): e1704.

15 C. Alvarado-Esquivel, 'Miscarriage history and Toxoplasma gondii infection: a cross-sectional study in women in Durango City, Mexico', *European Journal of Microbiology and Immunology* (2014); 4(2): 117–122; and F. Roberts, 'Histopathological features of ocular toxoplasmosis in the fetus and infant', *Archives of Ophthalmology* (2001); 119(1): 51–58

16 https://www.nhs.uk/conditions/pregnancy-and-baby/foods-to-avoid-pregnant/ (23 January 2017)

17 D. L. Villazanakretzer, 'Fish parasites: a growing concern during pregnancy', *Obstetrical & Gynecological Survey* (2016); 71(4): 253–259

18 C. M. Taylor, 'A review of guidance on fish consumption in pregnancy: is it fit for purpose?', *Public Health Nutrition* (2018); 21(11): 2149–2159

19 T. D. Solan, 'Mercury exposure in pregnancy: a review', *Journal of Perinatal Medicine* (2014); 42(6): 725–729

20 E. Ebel, 'Estimating the annual fraction of eggs contaminated with Salmonella enteritidis in the United States', *International Journal of Food Microbiology* (2000); 61(1): 51–62

21 A. Gyang, 'Salmonella Mississippi: a rare cause of second trimester miscarriage', *Archives of Gynecology and Obstetrics* (2008); 277(5): 437–438; K. Ravneet, 'A case of Salmonella typhi infection leading to miscarriage', *Journal of Laboratory Physicians* (2011); 3(1): 61–62; and S. E. Majowicz, 'The global burden of nontyphoidal salmonella gastroenteritis', *Clinical Infectious Diseases* (2010); 50(6): 882–889

22 https://www.bbc.co.uk/news/magazine-32033409 (25 March 2015)

23 A. Awofisayo, 'Pregnancy-associated listeriosis in England and Wales', *Epidemiology and Infection* (2015); 143(2): 249–256

24 M. Madjunkov, 'Listeriosis during pregnancy', *Archives of Gynecology and Obstetrics* (2017); 296(2): 143–152

25 https://www.cdc.gov/listeria/technical.html (12 December 2016)

26 Maggie Fox, 'Prepared salads recalled for salmonella, listeria risk', *NBC News* (19 October 2018)

27 M. Withers, 'Traditional beliefs and practices in pregnancy, childbirth and postpartum: a review of the evidence from Asian countries', *Midwifery* (2018); 56: 158–170

28 C. Nagata, 'Hot–cold foods in diet and all-cause mortality in a Japanese community: the Takayama study', *Annals of Epidemiology* (2017); 27(3): 194–199

29 O. Koren, 'Host remodeling of the gut microbiome and metabolic changes during pregnancy', *Cell* (2012); 150(3): 470–480; and A. N. Thornburn, 'Evidence that asthma is a developmental origin disease influenced by maternal diet and bacterial metabolites', *Nature Communications* (2015); 6: 7320

15: THE ALLERGY EPIDEMIC

1 https://www.cdc.gov/healthcommunication/toolstemplates/entertainmented/tips/Allergies.html (12 August 2019)

2 R. S. Gupta, 'Prevalence and severity of food allergies among US adults', *JAMA Netw Open* (2019); 2(1): e185630

3 Shayla Love, 'Food intolerance tests are shoddy science and traps for disordered eating', *Vice* (23 February 2018)

4 L. Wenyin, 'The epidemiology of food allergy in the global context', *International Journal of Environmental Research and Public Health* (2018); 15(9): 2043

5 C. Hammond, 'Unproven diagnostic tests for food allergy', *Immunology and Allergy Clinics of North America* (2018); 31(1): 153–163

6 D. Venkataram, 'Prevalence and longitudinal trends of food allergy during childhood and adolescence: results of the Isle of Wight Birth Cohort study', *Clinical and Experimental Allergy* (2018); 48(4): 394–402

7 E. Yousef, 'Clinical utility of serum specific IgE food testing in general practice: a tertiary care experience', *Journal of Allergy and Clinical Immunology* (2019); 143(2): AB275

8 B. P. Vickery, 'AR101 oral immunotherapy for peanut allergy', *New England Journal of Medicine* (2018); 379(21): 1991–2001

9 R. A. Pretorius, 'Maternal fiber dietary intakes during pregnancy and infant allergic disease', *Nutrients* (2019); 11(8): 1767

10 P. A. Eigenmann, 'Are avoidance diets still warranted in children with atopic dermatitis?', *Pediatric Allergy and Immunology* (2020); 1: 19–26

16: THE GLUTEN-FREE FAD

1 B. Lebwohl, 'Long term gluten consumption in adults without celiac disease and risk of coronary heart disease: prospective cohort study', *BMJ* (2017); 357: j1892

2 U. Volta, 'High prevalence of celiac disease in Italian general population', *Digestive Diseases and Science* (2011); 46(7): 1500–1505

3 J. R. Biesiekierski, 'Non-coeliac gluten sensitivity: piecing the puzzle together', *United European Gastroenterology* (2015); 3(2): 160–165

4 V. Melini, 'Gluten-free diet: gaps and needs for a healthier diet', *Nutrients* (2019); 11(1): 170

5 C. S. Johnston, 'Commercially available gluten-free pastas elevate postprandial glycemia in comparison to conventional wheat pasta in healthy adults: a double-blind randomized crossover trial', *Food Funct* (2017); 8(9): 3139–3144

6 I. D. Croall, 'Gluten does not induce gastrointestinal symptoms in healthy volunteers: a double-blind randomized placebo trial', *Gastroenterology* (2019); 157: 881–883

7 H. M. Roager, 'Whole grain-rich diet reduces body weight and systemic low-grade inflammation without inducing major changes of the gut microbiome: a randomised cross-over trial', *Gut* (2019); 68: 83–93

17: ON YOUR BIKE

1 UK exercise guidelines: https://www.nhs.uk/live-well/exercise/ (30 May 2018); US exercise guidelines: https://health.gov/paguidelines/ (2019)

2 W. W. Tigbe, 'Time spent in sedentary posture is associated with waist circumference and cardiovascular risk', *International Journal of Obesity* (2017); 41(5): 689–696

3 H. Fujita, 'Physical activity earlier in life is inversely associated with insulin resistance among adults in Japan', *Journal of Epidemiology* (2019); 29(2): 57–60

4 H. Pontzer, 'Hunter-gatherer energetics and human obesity', *PLOS ONE* (2012); 7(7): e40503

5 N. Casanova, 'Metabolic adaptations during negative energy balance and potential impact on appetite and food intake', *Proceedings of the Nutrition Society* (2019); 78(3): 279–289

6 D. M. Thomas, 'Why do individuals not lose more weight from an exercise intervention at a defined dose? An energy balance analysis', *Obesity Reviews* (2013); 13(10): 835–847

7 Alexi Mostrous, 'Coca-Cola spends £10m to counter links with obesity', *The Times* (18 December 2015); and Jonathan Gornall, 'Sugar: spinning a web of influence', *BMJ* (2015); 350: h231

8 M. Nestle, *Unsavory Truth: How Food Companies Skew the Science of What We Eat*, Basic Books (2018)

9 T. D. Noakes, 'Lobbyists for the sports drink industry: example of the rise of "contrarianism" in modern scientific debate', *Br J of Sports Med* (2007); 41(2): 107–109

10 L. M. Burke, 'Swifter, higher, stronger: What's on the menu?', *Science* (2018); 362(6416): 781–787

11 S. R. Chekroud, 'Association between physical exercise and mental health in 1.2 million individuals in the USA between 2011 and 2015', *Lancet Psychiatry* (2018); 5: 739–746

12 C. R. Gustafson, 'Exercise and the timing of snack choice: healthy snack choice is reduced in the post-exercise state', *Nutrients* (2018); 10(12): 1941

18: FOOD FOR THOUGHT

1 E. Jakubovski, 'Systematic review and meta-analysis: dose-response relationship of selective-serotonin reuptake inhibitors in major depressive disorder', *American Journal of Psychiatry* (2016); 173(2): 174–183

2 J. S. Lai, 'A systematic review and meta-analysis of dietary patterns and depression in community-dwelling adults', *American Journal of Clinical Nutrition* (2014); 99(1): 181–197; and D. Recchia, 'Associations between long-term adherence to healthy diet and recurrent depressive symptoms in Whitehall II Study', *European Journal of Nutrition* (2019); 1: 1–11

3 C. F. Reynolds, 'Early intervention to preempt major depression in older black and white adults', *Psychiatric Services* (2014); 65(6): 765–773

4 F. N. Jacka, 'A randomised controlled trial of dietary improvement for adults with major depression (the "SMILES" trial)', *BMC Medicine* (2017); 15(1): 23

5 J. Firth, 'The effects of dietary improvement on symptoms of depression and anxiety: a meta-analysis of randomized controlled trials', *Psychosomatic Medicine* (2019); 81(3): 265–280; and S. Mizuno, 'Bifidobacterium-rich fecal donor may be a positive predictor for successful fecal microbiota transplantation in patients with irritable bowel syndrome', *Digestion* (2017); 96(1): 29–38

6 A. Sánchez-Villegas, 'Mediterranean dietary pattern and depression: the PREDIMED randomized trial', *BMC Medicine* (2013); 11: 208

7 M. Valles Colomer, 'The neuroactive potential of human gut microbiota in quality of life and depression,' *Nature Microbiology* (2019); 4: 623–632

8 J. M. Yano, 'Indigenous bacteria from the gut microbiota regulate host serotonin biosynthesis', *Cell* (2015); 161(2): 264–276

9 I. Lukić, 'Antidepressants affect gut microbiota and Ruminococcus flavefaciens is able to abolish their effects on depressive-like behavior', *Translational Psychiatry* (2019); 9(1): 133

10 M. J. Walters, 'Associations of lifestyle and vascular risk factors with Alzheimer's brain biomarkers during middle age', *BMJ OPEN* (2018); 8(11): e023664

11 T. Akbaraly, 'Association of long-term diet quality with hippocampal volume: longitudinal cohort study', *American Journal of Medicine* (2018); 131(11): 1372–1381

12 S. E. Setti, 'Alterations in hippocampal activity and Alzheimer's disease', *Translational Issues in Psychological Science* (2018); 3(4): 348–356

13 P. Zheng, 'The gut microbiome from patients with schizophrenia modulates the glutamate-glutamine-GABA cycle and schizophrenia-relevant behaviors in mice', *Science Advances* (2019); 5(2): eaau8317

14 I. Argou-Cardozo, 'Clostridium bacteria and autism spectrum conditions: a systematic review and hypothetical contribution of environmental glyphosate Levels', *Medical Sciences* (2018); 6(2): 29

15 D. W. Kang, 'Differences in fecal microbial metabolites and microbiota of children with autism spectrum disorders', *Anaerobe* (2018); 49: 121–131

16 S. Mizuno, 'Bifidobacterium-rich fecal donor may be a positive predictor for successful fecal microbiota transplantation in patients with irritable bowel syndrome', *Digestion* (2017); 96(1): 29–38

17 M. I. Butler, 'From isoniazid to psychobiotics: the gut microbiome as a new antidepressant target', *British Journal of Hospital Medicine* (2019); 80(3): 139–145

18 F. N. Jacka, 'Maternal and early postnatal nutrition and mental health of offspring by age 5 years: a prospective cohort study', *J Acad Child & Adol Psych* (2013); 52(10): 1038–1047

19 Felice Jacka, *Brain Changer: How diet can save your mental health*, Yellow Kite (2019)

19: THE DIRTY BUSINESS OF WATER

1 A. Saylor, 'What's wrong with the tap? Examining perceptions of tap water and bottled water at Purdue University', *Environmental Management* (2011); 48(3): 588–601

2 D. Lantagne, 'Household water treatment and cholera control', *Journal of Infectious Diseases* (2018); 218(3): s147–s153

3 M. McCartney, 'Waterlogged?', *BMJ* (2011); 343: d4280

4 F. Rosario-Ortiz, 'How do you like your tap water?', *Science* (2016); 351(6267): 912–914

5 E. Brezina, 'Investigation and risk evaluation of the occurrence of carbamazepine, oxcarbazepine, their human metabolites and transformation products in the urban water cycle', *Environmental Pollution* (2017); 225: 261–269

6 T. Spector, *Identically Different*, Weidenfeld & Nicolson (2012)

7 M. Wagner, 'Identification of putative steroid receptor antagonists in bottled water', *PLOS ONE* (2013); 8(8): e72472

8 W. Huo, 'Maternal urinary bisphenol A levels and infant low birth weight: a nested case-control study of the Health Baby Cohort in China', *Environmental International* (2015); 85: 96–103; and H. Gao, 'Bisphenol A and hormone-associated cancers: current progress and perspectives', *Medicine* (2015); 94(1): e211

9 EFSA, 'Bisphenol A: new immune system evidence useful but limited', *EFSA Reports* (13 October 2016)

10 Z. Iheozor-Ejiofor, 'Water fluoridation for the prevention of dental caries', *Cochrane Database of System Reviews* (2015); 6: CD010856

11 J. R. Jambeck, 'Marine pollution. Plastic waste inputs from land into the ocean', *Science* (2015) 13; 347(6223): 768–71

12 P. G. Ryan, 'Monitoring the abundance of plastic debris in the marine environment', *Proceedings Transactions Royal Soc B* (2009); 364: 1999–2012

13 L. M. Bartoshuk, 'NaCl thresholds in man: thresholds for water taste or NaCl taste?', *Journal of Comparative and Physiological Psychology* (1974); 87(2): 310–325

20: JUST A DROP

1 https://www.alcohol.org/guides/global-drinking-demographics/ (2019)

2 D. W. Lachenmeier, 'Comparative risk assessment of alcohol, tobacco, cannabis and other illicit drugs using the margin of exposure approach', *Scientific Reports* (2015); 5: 8126

3 R. Bruha, 'Alcoholic liver disease', *World Journal of Hepatology* (2012); 4(3): 81–90; and G. P. Jordaan, 'Alcohol-induced psychotic disorder: a review', *Metabolic Brain Disease* (2104); 29(2): 231–243

4 https://www.alcohol.org/guides/global-drinking-demographics/ (2019)

5 A. S. St Leger, 'Factors associated with cardiac mortality in developed countries with particular reference to the consumption of wine', *Lancet* (1979); 1(8124): 1017–1020; and A. Di Castelnuovo, 'Alcohol dosing and total mortality in men and women: an updated meta-analysis', *Archives of Internal Medicine* (2006); 166(22): 2437–2445

6 https://www.gov.uk/government/news/new-alcohol-guidelines-show-increased-risk-of-cancer (8 January 2016)

7 B. Xi, 'Relationship of alcohol consumption to all-cause, cardiovascular, and cancer-related mortality in US adults', *J. American College of Cardiology* (2017); 70(8): 913–922

8 K. A. Welch, 'Alcohol consumption and brain health', *BMJ* (2017); 357: j2645

9 S. Sabia, 'Alcohol consumption and risk of dementia: 23 year follow-up of Whitehall II cohort study', *BMJ* (2018); 362: k2927

10 J. Holt-Lunstad, 'Social relationships and mortality risk: a meta-analytic review', *PLOS Medicine* (2010); 7(7): e1000316

11 A. M. Wood, 'Risk thresholds for alcohol consumption: combined analysis of individual-participant data for 599,912 current drinkers in 83 prospective studies', *The Lancet* (2018); 391(10129): 1513–1523

12 M. G. Griswold, 'Alcohol use and burden for 195 countries and territories, 1990–2016: a systematic analysis for the Global Burden of Disease Study 2016', *The Lancet* (2018); 392(10152): 1015–1035

13 A. L. Freeman, 'Communicating health risks in science publications: time for everyone to take responsibility', *BMC Medicine* (2018); 16(1): 207

14 H. J. Edenberg, 'The genetics of alcohol metabolism: role of alcohol dehydrogenase and aldehyde dehydrogenase variants', *Alcohol Research and Health* (2007); 30(1): 5–13

15 S. M. Ruiz, 'Closing the gender gap: the case for gender-specific alcoholism research', *Journal of Alcoholism and Drug Dependence* (2013); 1(6): e106

16 V. Vatsalya, 'A review on the sex differences in organ and system pathology with alcohol drinking', *Current Drug Abuse Reviews* (2017); 9(2): 87–92

17 Peter Lloyd, 'Deadly link between alcohol and breast cancer is "ignored by middle-aged women who are *most* at risk of developing the disease"', *Mail Online* (13 February 2019)

18 M. I. Queipo-Ortuño, 'Influence of red wine polyphenols and ethanol on the gut microbiota ecology and biomarkers', *Am Journal of Clinical Nutrition* (2012); 95(6): 1323–1334

19 A. Chaplin, 'Resveratrol, metabolic syndrome, and gut microbiota', *Nutrients* (2018); 10(11): e1651; and X. Fan, 'Drinking alcohol is associated with variation in the human oral microbiome in a large study of American adults', *Microbiome* (2018); 6(1): 59

20 C. I. LeRoy, 'Red wine consumption associated with increased gut microbiota α-diversity in 3 independent cohorts', *Gastroenterology* (2019); pii: S0016–5085(19): 41244–4

21 R. O. de Visser, 'The growth of "Dry January": promoting participation and the benefits of participation', *Eur J Public Health* (2017); 27(5): 929–931

22 T. S. Naimi, 'Erosion of state alcohol excise taxes in the United States', *Journal of Studies on Alcohol and Drugs* (2018); 79(1): 43–48

23 https://www.cdc.gov/alcohol/index.htm (2019)

24 Z. Zupan, 'Erosion of state alcohol excise taxes in the United States', *BMJ* (2017); 359: j5623

21: FOOD MILES

1 D. Coley, 'Local food, food miles and carbon emissions: a comparison of farm shop and mass distribution approaches', *Food Policy* (2009); 34(2): 150–155

2 C. Saunders, 'Food miles, carbon footprinting and their potential impact on trade', *Semantic Scholar* (2009); AARES 53rd annual conference at Cairns, 10–13 February 2009

3 E. Soode-Schimonsky, 'Product environmental footprint of strawberries: case studies in Estonia and Germany', *J Environ Management* (2017); 203(Pt 1): 564–577

4 W. Willett, 'Food in the Anthropocene: the EAT-Lancet Commission on healthy diets from sustainable food systems', *The Lancet* (2019); 393(10170): 447–492

5 J. Milner, 'Health effects of adopting low greenhouse gas emission diets in the UK', *BMJ Open* (2015); 5: e007364

6 J. Poore, 'Reducing food's environmental impacts through producers and consumers', *Science* (2018); 360: 987–992

7 T. D. Searchinger, 'Assessing the efficiency of changes in land use for mitigating climate change', *Nature* (2018); 564: 249–253

8 George Monbiot, 'We can't keep eating as we are – why isn't the IPCC shouting this from the rooftops?', *The Guardian* (9 August 2019)

22: SPRAYING THE PLANET

1 R. Mesnage, 'Facts and fallacies in the debate on glyphosate toxicity', *Frontiers in Public Health* (2017); 5: 316

2 https://www.iarc.fr/wp-content/uploads/2018/07/MonographVolume112-1.pdf (20 March 2015)

3 Ben Webster, 'Weedkiller scientist was paid £120,000 by cancer lawyers', *The Times* (18 October 2017)

4 P. J. Mills, 'Excretion of the herbicide glyphosate in older adults between 1993 and 2016', *JAMA* (2017); 318(16): 1610–1611

5 J. V. Tarazona, 'Glyphosate toxicity and carcinogenicity: a review of the scientific basis of the European Union assessment and its differences with IARC', *Archives of Toxicology* (2017); 91(8): 2723–2743; and C. J. Portier, 'Update to Tarazona et al. (2017): glyphosate toxicity and carcinogenicity: a review of the scientific basis of the European Union assessment and its differences with IARC', *Archives of Toxicology* (2018); 92(3): 1341

6 E. T. Chang, 'Systematic review and meta-analysis of glyphosate exposure and risk of lymphohematopoietic cancers', *Journal of Environmental Science and Health, Part B* (2016); 51(6): 402–434

7 C. Gillezeau, 'The evidence of human exposure to glyphosate: a review', *Environmental Health* (2019); 18(1): 2; and M. E. Leon, 'Pesticide use and risk of non-Hodgkin lymphoid malignancies in agricultural cohorts from France, Norway and the USA: a pooled analysis from the AGRICOH consortium', *International Journal of Epidemiology* (2019); 48(5): 1519–1535

8 L. Hu, 'The association between non-Hodgkin lymphoma and organophosphate pesticides exposure: a meta-analysis', *Environmental Pollution* (2017); 231: 319–328

9 B. González-Alzaga, 'A systematic review of neurodevelopmental effects of prenatal and postnatal organophosphate pesticide exposure', *Toxicology Letters* (2014); 230(2): 104–121; and Y. Chiu, 'Association between pesticide residue intake from consumption of fruits and vegetables and pregnancy outcomes among women undergoing infertility treatment with assisted reproductive technology', *JAMA* (2018); 178(1): 17–26

10 F. Manservisi, 'The Ramazzini Institute 13-week pilot study glyphosate-based herbicides administered at human-equivalent dose to Sprague Dawley rats', *Environmental Health* (2019); 18(1): 15; and Y. Aitbali, 'Glyphosate-based herbicide exposure affects gut microbiota, anxiety and depression-like behaviors in mice', *Neurotoxicology and Teratology* (2018); 67: 44–49

11 E. V. Motta, 'Glyphosate perturbs the gut microbiota of honey bees', *PNAS* (2018); 115(41): 10305–10310

12 J. Baudry, 'Association of frequency of organic food consumption with cancer risk: findings from NutriNet-Santé Prospective Cohort Study', *JAMA* (2018); 178(12): 1597–1606

13 K. E. Bradbury, 'Organic food consumption and the incidence of cancer in a large prospective study of women in the UK', *British Journal of Cancer* (2014); 110: 2321–2326

14 http://www.anh-usa.org/wp-content/uploads/2016/04/ANHUSA-glyphosate-breakfast-study-FINAL.pdf (19 April 2016)

23: DON'T TRUST ME, I'M A DOCTOR

1 K. Womersley, 'Medical schools should be prioritising nutrition and lifestyle education', *BMJ* (2017); 359: j4861

2 J. Crowley, 'Nutrition in medical education: a systematic review', *Lancet Planetary Health* (2019); 9: PE379–E389

3 S. Greenhalgh, 'Making China safe for Coke: how Coca-Cola shaped obesity science and policy in China', *BMJ* (2019); 364: k5050

4 M. E. Lean, 'Primary care-led weight management for remission of type 2 diabetes (DiRECT): an open-label, cluster-randomised trial', *The Lancet* (2018); 391(10120): 541–551

5 https://www.ncbi.nlm.nih.gov/pubmed/21366836; D. Zhu, 'The relationship between health professionals' weight status and attitudes towards weight management: a systematic review', *Obesity Reviews* (2011); 12(5): e324–337

6 nutritank.com and thedoctorskitchen.com

7 K. E. Aspry, 'Medical nutrition education, training and competencies to advance guideline-based diet counseling by physicians', *Circulation* (2018); 137: e821–e841

CONCLUSION: HOW TO EAT

1 D. McDonald, 'American gut: an open platform for citizen science microbiome research', *mSystems* (2018); 3(3): e00031–18

2 joinzoe.com

3 M. J. Blaser, 'Antibiotic use and its consequences for the normal microbiome', *Science* (2016); 352: 544–545

4 R. de Cabo, 'Effects of intermittent fasting on health, aging and disease', *New England Journal of medicine* (2019); 381: 2541–51

5 US Burden of Disease Collaborators, 'The state of us health, 1990–2010: burden of diseases, injuries, and risk factors', *JAMA* (2013); 310(6): 591–606

6 Laura Reiley, 'How the Trump administration limited the scope of the USDA's 2020 dietary guidelines', *Washington Post* (30 August 2019)

7 https://www.nationalfoodstrategy.org/the-report/

8 Ron Sterk, ' EU Sugar producers suffer after reform', *Food Business News* (8 August 2019)

9 H. Moses, 'The anatomy of medical research: US and international comparisons', *JAMA* (2015); 313(2): 174–89

10 R. G. Kyle, 'Obesity prevalence among healthcare professionals in England: a cross-sectional study using the Health Survey for England', *BMJ Open* (2017); 4 Dec: 018498; and S. E. Luckhaupt, 'Prevalence of obesity among US workers and associations with occupational factors', *Am J Prev Med* (2014); 46(3): 237–248

INDEX

Now you know what *not* to believe, but what *should* we all know about nutrition?

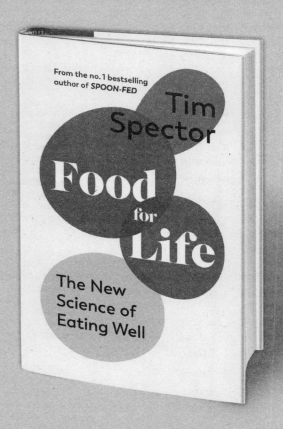

From the no. 1 bestselling author of *SPOON-FED*

Tim Spector

Food for Life

The New Science of Eating Well

A comprehensive guide to the new science of nutrition from the #1 bestselling author of *Spoon-Fed*

COMING OCTOBER 2022